Lecture Notes in Computer Science

Edited by G. Goos, J. Hartmanis and J. van Leeuwen

Advisory Board: W. Brauer   D. Gries   J. Stoer

Jonathan S. Greenfield

Distributed Programming Paradigms with Cryptography Applications

Springer-Verlag

Berlin Heidelberg New York
London Paris Tokyo
Hong Kong Barcelona
Budapest

Series Editors

Gerhard Goos
Universität Karlsruhe
Postfach 69 80, Vincenz-Priessnitz-Straße 1, D-76131 Karlsruhe, Germany

Juris Hartmanis
Department of Computer Science, Cornell University
4130 Upson Hall, Ithaka, NY 14853, USA

Jan van Leeuwen
Department of Computer Science, Utrecht University
Padualaan 14, 3584 CH Utrecht, The Netherlands

Author

Jonathan S. Greenfield
Distributed Computing Environments Group, M/S B272
Los Alamos National Laboratory
Los Alamos, New Mexico 87545, USA

CR Subject Classification (1991): D.1.3, D.2.m, E.3, G.3

ISBN 3-540-58496-X Springer-Verlag Berlin Heidelberg New York

CIP data applied for

© Springer-Verlag Berlin Heidelberg 1994
Printed in Germany

Typesetting: Camera-ready by author
SPIN: 10479104   45/3140-543210 - Printed on acid-free paper

Foreword

Jonathan Greenfield's thesis work in distributed computing, presented in this monograph, is part of my research project on programming paradigms for parallel scientific computing at Syracuse University. The starting point of this research is the definition of a *programming paradigm* as "a class of algorithms that solve different problems but have the same control structure." Using this concept, two or more parallel programs can be developed simultaneously.

First, you write a *parallel generic program* that implements the common control structure. The generic program includes a few unspecified data types and procedures that vary from one application to another. A *parallel application program* is obtained by replacing these types and procedures with the corresponding ones from a sequential program that solves a particular problem. The clear separation of the issues of parallelism from the details of applications leads to parallel programs that are easy to understand.

Jonathan Greenfield has convincingly demonstrated the power of this programming methodology in this monograph on *Distributed Programming Paradigms with Cryptography Applications*. The work makes several contributions to parallel programming and cryptography:

1. The monograph describes five *parallel algorithms* for the *RSA cryptosystem*: enciphering, deciphering, modulus factorization, and two methods for generating large primes.

2. The *parallel performance* of these algorithms was tested on a reconfigurable *multicomputer*—a Computing Surface with 48 transputer processors and distributed memory.

3. *RSA enciphering* and *deciphering* are very time-consuming operations that traditionally use a serial method (cipher block chaining) to break long messages into shorter blocks. Jonathan has invented a new method, called *message block chaining*, which permits fast enciphering and deciphering of message blocks on a parallel computer.

4. The RSA cryptosystem requires the generation of large primes. The primality of large random integers has traditionally been certified by the probabilistic Miller-Rabin algorithm. Jonathan has proposed a new algorithm that uses *deterministic primality testing* for a particular class of large random integers.

5. The five distributed programs are derived as instances of two *programming paradigms* invented by Jonathan: the *synchronized servers* and the *competing servers* paradigms. The latter is particularly interesting, since it demonstrates that non-determinism is necessary for efficient use of parallel computing in some user applications—which was an unexpected surprise for me.

The work is an appealing combination of the *theory* and *practice* of parallel computing. From the point of view of a computer user, the most important contribution is the new method of fast RSA enciphering and deciphering, which can be implemented in hardware. From the point of view of a computer scientist, it is an amazing feat to recognize five different aspects of the same application as instances of two simple programming paradigms for parallel computing. In addition, the monograph is well-written and easy to understand.

June 1994   Per Brinch Hansen

Acknowledgments

"When a true genius appears in this world you may know him by the sign that the dunces are all in confederacy against him."

Jonathan Swift

This monograph is a revised version of my doctoral thesis, prepared under the direction of Per Brinch Hansen at Syracuse University. I owe many thanks to Professor Brinch Hansen who introduced me to programming paradigms and to the problem of parallel primality testing. The latter sparked my interest in cryptography, while the former gave me a direction for my research.

Per Brinch Hansen has kindly granted permission to include revised versions of his occam program of June 9, 1992 for primality testing with multiple-length arithmetic.

In addition, I am grateful to Professor Brinch Hansen for the enormous amount of time he spent critiquing earlier versions of this monograph. His many useful suggestions have improved this work considerably. I must also thank Geoffrey Fox, David Gries, Salim Hariri, Carlos Hartmann, Nancy McCracken and Ernest Sibert for their helpful comments.

Finally, I thank my wife Georgette, my sons Channon and Alexzander, and especially my parents Judith and Stuart, for their support and toleration.

June 1994   Jonathan Greenfield

Table of Contents

Chapter 1

Overview

"Obviously, the highest type of efficiency is that which can utilize existing material to the best advantage."

Jawaharlal Nehru

1.1 Background

More than three years ago, Per Brinch Hansen began a research project exploring the role of programming paradigms in distributed computing [Brinch Hansen 90a-c, 91a-e, 92a-f, 93]. As part of that work, he developed a distributed algorithm for primality testing, and introduced me to parallel primality testing [Brinch Hansen 92b,c].

After examining the problem of primality testing, I became interested in the more complex problem of prime generation, and eventually in the problem of generating strong primes for use in constructing keys for the RSA cryptosystem. Once I had become familiar with the RSA cryptosystem, I became interested in other related problems, including RSA enciphering and deciphering, and factorization.

Developing a distributed algorithm for strong prime generation turned out to be a more difficult task than I had expected. Once I had developed an efficient distributed program, I realized that the technique I had used to parallelize the generation of primes could be applied to a number of other problems. With this realization, my emphasis began to move in the direction of Brinch Hansen's work with programming paradigms.

A *programming paradigm* is "a class of algorithms that solve different problems but have the same control structure" [Brinch Hansen 93]. A *program shell* (or *generic program*) implements the control structures of a programming paradigm, but does not implement specific computations for any particular application. Subsequently, the program shell can be used to implement any algorithm that fits the control structure of the paradigm. Implementation of a particular algorithm is done by simple substitution of algorithm-specific data types and procedures.

A *distributed program shell* implements a programming paradigm for execution on a distributed computer. Therefore, a distributed program shell can be adapted to implement a distributed implementation of any algorithm that fits the control structure of the paradigm. The program shell is adapted by simple substitution of data types and sequential procedures.

My work over the last year and a half has explored the development of distributed program shells [Chapters 4, 7] and their use in implementing several problems related to RSA cryptography [Chapters 5, 6, 8, 9]. This chapter provides an overview of that work, along with a discussion of my goals and the results of the work.

Related work in the area of parallel programming paradigms has been undertaken by Brinch Hansen [90a-c, 91a-e, 92a-f, 93], Cole [89], Nelson [87], Rao [91] and Stanford-Clark [91].

1.2 Public-Key Cryptography and RSA

A cryptosystem is used by a *cryptographer* to *encipher* (or *encrypt*) a *plaintext* message to produce an unintelligible *cryptogram* (or *ciphertext* message). An authorized receiver should be able to *decipher* the cryptogram to recover the original plaintext message. An unauthorized receiver (*cryptanalyst*), however, must be unable to *decrypt* the ciphertext. (Note that decryption is *unauthorized* deciphering.) [Brassard 88]

In 1976, Diffie and Hellman introduced the idea of a *public-key cryptosystem* [Diffie 76]. Subsequently, Rivest, Shamir and Adleman proposed the first example of a public-key cryptosystem, now known as the *RSA cryptosystem*. The cryptosystem is based on modular exponentiation with a fixed exponent and modulus [Rivest 78b].

Public-key cryptosystems have one main advantage over secret-key cryptosystems: the absence of a *key distribution problem* [Welsh 88]. With conventional secret-key cryptography, a single key is used for both enciphering and deciphering. For a sender and receiver to communicate using a secret-key cryptosystem, both must possess the secret key. This leads to the problem of finding a secure way to communicate the secret-key to those who need to use it.

Public-key cryptosystems solve this problem by using two separate keys for enciphering and deciphering. The enciphering key is made public, while the deciphering key is kept secret. Since the enciphering key is public, anybody can encipher a message using the cryptosystem. However, only an authorized receiver who possesses the secret deciphering key is able to decipher the messages. Consequently, the public enciphering key can be distributed without security concerns.

Despite this advantage, the RSA system has been slow to gain acceptance for practical use. This is primarily due to the large computational requirements of RSA enciphering and deciphering.

Since speed is the main barrier to the use of RSA in practical systems, numerous simplifications and optimizations have been proposed in order improve the speed of RSA enciphering and deciphering [Agnew 88, Beth 86, Bos 89, Cooper 90, Findlay

89, Geiselmann 90, Ghafoor 89, Jung 87, Kawamura 88, Morita 89, Yacobi 90].

1.3 Research Goals

The original goal of my work was to explore distributed computing as a tool for speeding up the basic RSA operations of key generation, enciphering and deciphering. To this point in time, the role of public-key cryptography in practical applications has been primarily limited to that of a key exchange mechanism in *hybrid* cryptosystems [Beth 91]. A hybrid cryptosystem uses public-key cryptography to securely communicate a secret key. The secret key is subsequently used for conventional secret-key cryptography. My goal was to determine whether distributed computing could make public-key cryptography practical for use without a hybrid approach.

It has never been my goal, however, to produce the world's fastest implementations of these public-key cryptography applications. As a practical matter, it could not be. The hardware I had at my disposal, a *Meiko Computing Surface* [Meiko 88] with T800 *transputers* [INMOS 88a], required a software implementation of the *multiple-length arithmetic* necessary for RSA applications.

To implement multiple-length arithmetic, I used low-level procedures provided by *occam 2* [INMOS 88b] to manipulate base-2^{32} digits. While this resulted in an extremely compact representation and fast arithmetic, my software implementations could not compete with special-purpose hardware implementations of multiple-length arithmetic [Barrett 86, Beth 91, Brickell 89, Er 91, Hoornaert 88, Orton 86, Orup 90, Rankine 86, Rivest 80]. Since developing hardware implementations was not an option, my goal has been to develop distributed programs that serve as models for distributed hardware implementations.

In addition, I have sought to explore the use of programming paradigms as a tool for developing distributed programs. Brinch Hansen [93] has observed that the process of identifying a programming paradigm and selecting two interesting algorithms that fit the paradigm is not an easy one. One goal of my work has been to further explore the feasibility of developing distributed programs using previously-developed distributed program shells.

1.4 Prime Generation

An integer $p > 1$ is *prime* if its only factors are 1 and p. An integer $n > 1$ is *composite* if it is the product of more than one prime.

Generating large primes is an important problem in RSA cryptography because an RSA modulus is constructed as the product of two large primes [Rivest 78b]. Typically, an RSA modulus must be an integer with one hundred and fifty to three hundred decimal digits [Beth 91, Brassard 88]. This corresponds to primes of seventy-five to one hundred and fifty digits, each.

The conventional method for testing integers of this size for primality is a test known as the *Miller-Rabin witness test* [Miller 76, Rabin 80]. The Miller-Rabin test is a probabilistic test for compositeness. The test either proves that the given number is composite, or it proves nothing.

Formally, Rabin has proved that the probability a composite will go undetected by this test is no more than 0.25. In practice, it has been observed that the chance that a composite number goes undetected by the test is very close to zero [Beauchemin 86, Gordon 84, Jung 87, Rivest 90].

To test a number for primality, the Miller-Rabin test is simply repeated some fixed number of times m, to make the probability of an error negligible [Beauchemin 86, Gordon 84, Jung 87]. We select $m = 40$, consistent with [Brinch Hansen 92b]. This value of m insures that there will be no practical chance that the Miller-Rabin test will fail. However, a formal bound on the probability of an error is rather difficult to compute and depends upon the size of the number being tested for primality [Beauchemin 86].

When Brinch Hansen first showed me his distributed implementation of primality testing, I brazenly suggested that the program could be easily extended to efficiently *generate* a prime number. Brinch Hansen suggested that I implement the changes.

Since a single Miller-Rabin witness test is such a good indicator of primality, I planned to extend the distributed program by allowing individual server processes to test distinct numbers for primality, until some number was found that passed the single witness test. Such a number would very likely be prime, and would be called a *prime candidate*. Once a prime candidate had been found, the candidate would be certified in parallel using Brinch Hansen's algorithm. Since each Miller-Rabin test is fairly slow [Brinch Hansen 92b], this would save a substantial amount of time compared to testing each number by a complete certification test.

Upon examining the literature, I found that this was a standard technique for generating primes [Jung 87, Rivest 90]. I also found that another technique was additionally used to reduce execution time even further. Since *trial division* by small primes (for example, primes less than 500) is much faster than a single Miller-Rabin test, even more time can be saved by first testing a number for small prime factors [Jung 87].

This leads to a two-tiered test for prime candidates. An arbitrary number is first tested for small prime factors using trial division. If a small prime factor is found, the number is discarded. Otherwise, a single Miller-Rabin test is performed. A number that passes both tests is a prime candidate that warrants the full Miller-Rabin certification procedure.

With this addition of trial division, the modifications turned out to be far more complicated than I had expected. The basic problem was one of *load balance*. Prime numbers are rather haphazardly distributed among the integers [Bressoud 89, Riesel 85]. The same is true of numbers without small prime factors. Consequently, while one process tests a number without small factors, which requires use of the slow Miller-Rabin test, another process might quickly find a small prime factor in the number it tests.

To produce a reasonably efficient parallel program, it was necessary to allow each server process to search at its own rate, independent of the other server processes. Implementing this required *non-deterministic communication* [Chapter 5] and resulted in the development of the *competing servers* paradigm [Chapter 4].

Finding an application that required non-deterministic communication to be efficient was an interesting result, by itself. After studying a number of regular computations, Brinch Hansen [93] had adopted the tentative position that non-deterministic communication was necessary for system-level programming, but not for application-level programming. Parallel prime generation demonstrates that, at least in some cases, non-determinism can be important to application-level programs. This conclusion is consistent with the hypothesis of Yelick [93] that applications with some irregularity in their structure require non-determinism.

1.5 The Competing Servers Array

The competing servers paradigm includes any problem that can be viewed as a search, and for which only one solution is required. In the case of generating a prime candidate, the problem can be considered a search for a prime candidate, of some specified size in digits. While many suitable candidates exist, the prime generation algorithm requires just one.

In general, a competing servers search need not be a search in the conventional sense. We also use competing servers to implement factoring. While factoring isn't a problem that would conventionally be thought of as a search problem, we can view it as a search for factors of a given product. Consequently, the competing servers paradigm includes a larger class of algorithms than the term "search" might imply.

I decided to implement a generic program for competing servers as a linear array, as shown in figure 1.1. I use squares (and later, circles) to represent processes, while directed edges represent communication channels. The direction of an arrow indicates the direction in which messages may be sent. An edge with arrows at both ends represents a pair of channels.

There were several reasons for the decision to use an array. Each transputer can be directly connected to at most four other transputers. This ruled out the possibility of having a single master connected to each server process. Of the possible configurations available, I settled on a linear array primarily because it was simple and because communication time was not critical, at least for my applications.

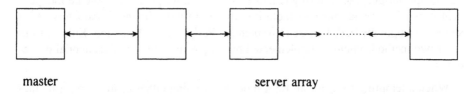

master   server array

Figure 1.1: Master and server array for the competing servers paradigm

Simplicity is a particularly important concern, as non-deterministic communication substantially complicates the program. The simple array topology keeps the program as simple as possible. Since the array implementation was sufficient for the two applications I was implementing, and was likely to be sufficient for many other applications, there was no reason to adopt a more complex topology.

1.6 Special-Purpose Factoring

As we mentioned earlier, an RSA modulus is constructed as the product of two large primes. This is done to prevent factorization of the modulus, as factoring the modulus breaks the cryptosystem [Rivest 78b, Chapter 2].

Factoring algorithms can be separated into two broad classes, *special-purpose factoring algorithms* and *general-purpose factoring algorithms*. A special-purpose factoring algorithm has an execution time that depends upon some special property of the product being factored. For example, an algorithm that factors products with small factors faster than products without small factors is a special-purpose algorithm. A general-purpose factoring algorithm has an execution time that depends solely upon the size of the product.

For an RSA modulus to be secure, it must not be possible to factor the modulus using any algorithm. Constructing the modulus as a product of two large primes renders general-purpose algorithms, as well as some special-purpose algorithms, useless [Brassard 88, Chapter 5].

The prime factors must possess additional properties, however, for the modulus to be secure against several other special-purpose factoring attacks [Gordon 84, Rivest 78a, Chapter 5]. One way to ensure that an RSA modulus is secure is to test it with a special-purpose factoring attack [Chapter 6]. If the special-purpose algorithms cannot factor the modulus, the modulus can be considered secure.

Several important special-purpose factoring algorithms are *parameterized* methods that use a probabilistic approach. Among these are *Pollard's* $p-1$ *method* [Pollard 74], *Williams'* $p+1$ *method* [Williams 82] and the *Elliptic Curve method* [Lenstra 87]. Each uses a parameter that may be arbitrarily varied. It has been shown that these parameterized methods are more likely to successfully factor a product when a number of short runs (each with different choice of parameter) are used, than when the same amount of execution time is spent on one long run (with a single a single choice of parameter) [Bressoud 89].

The second application of the competing servers array is a distributed implementation of special-purpose factoring [Chapter 6]. The server processes execute independent runs of a special-purpose factoring algorithm with distinct parameters. As a simple example method, I chose to implement Pollard's $p-1$ method. Either of the other two methods could be implemented by simple substitution of sequential procedures.

When attempting to factor an RSA modulus, finding either of the two prime factors reveals the complete prime factorization. Consequently, it is only necessary that a single factor be returned. This makes modulus factorization suitable for competing servers.

Since these parameterized factoring methods may fail to factor a given product, their execution is terminated if no factor is found before some pre-determined time limit [Bressoud 89]. A minor change to the master process of the generic competing servers program allows execution to be timed-out after a pre-determined time limit is reached [Chapter 4].

1.7 RSA Enciphering and Deciphering

A long message to be enciphered using the RSA cryptosystem must first be broken into small message blocks. The individual blocks are subsequently enciphered. A method used to break a message into blocks is known as a *mode of operation*.

When I first examined RSA enciphering, it seemed obvious to me that the small message blocks could be enciphered in parallel. Consequently, I was rather surprised when I did not find this approach mentioned in the literature. Some further research revealed that *cipher block chaining*, the mode of operation normally used for RSA enciphering [Brassard 88], prevented this parallel approach.

To remedy the serialization of cipher block chaining, I proposed a new mode of operation, *message block chaining*, which preserves many of the important properties of cipher block chaining, but allows the individual message blocks to be enciphered in parallel [Chapter 8]. I later discovered that my parallel approach had been proposed previously [Er 91, Yun 90]. However, these implementations had simply ignored the serialization problem of cipher block chaining by using a different mode of operation, *electronic code book*, with known faults [Brassard 88].

Once I had developed message block chaining, RSA enciphering and deciphering was a straightforward application of the *synchronized servers* pipeline [Chapter 7].

1.8 The Synchronized Servers Pipeline

The synchronized servers paradigm includes any problem where a particular computation is repeated with different input data. The synchronized servers pipeline may be used to implement any problem suitable for a simple *domain decomposition* or *data parallel* approach [Cok 91, Fox 88], provided that the computation time is large enough to make the communication time negligible.

The generic synchronized servers program uses a distributor process connected to a pipeline of servers, connected to a collector process, as shown in figure 1.2.

The only complicated parts of the generic program are the distribution of input data, and the output of computed results. Both input and output are implemented in shift-register fashion, with input data shifted into the pipeline, and computed results shifted out. The main difficulty implementing this input and output is handling cases where the total number of individual computations does not equal the number of server processes.

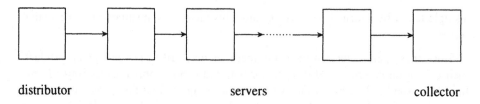

distributor   servers   collector

Figure 1.2: *Synchronized servers pipeline with distributor and collector*

1.9 Generating Deterministically Certified Primes

As we mentioned earlier, the probabilistic Miller-Rabin test is conventionally used to certify large primes. While nonprobabilistic certification algorithms do exist [Adleman 87, Goldwasser 86], they are impractically slow to use [Beauchemin 86].

Generating a prime is notably different from testing an arbitrary number for primality, however. An arbitrary number cannot be expected to have any special characteristics. When generating a prime, however, it is possible to generate only numbers with special characteristics. Correspondingly, a primality test that requires those special characteristics can be used to test those numbers.

Shawe-Taylor [86] and Maurer [89] describe algorithms to generate deterministically certified primes. To certify an integer n as prime, these certification algorithms require a partial factorization of $n-1$. Correspondingly, n is generated so that a suitable partial factorization of $n-1$ is known. Unfortunately, these algorithms are not amenable to a parallel implementation.

I propose a related algorithm to deterministically test an integer n for primality when all of the prime factors of $n-1$ are known. The algorithm is based on a theorem due to Edouard Lucas [Ribenboim 91]. Once prime candidate n has been generated so that all the prime factors of $n-1$ are known, the candidate can be deterministically tested for primality by performing a test with each prime factor of $n-1$.

The algorithm can be implemented as a parallel program by using both a competing servers array and a synchronized servers pipeline. Competing servers are used to generate prime candidate n such that the prime factors of $n-1$ are known. Subsequently, synchronized servers are used to test n with each of those prime factors.

1.10 Conclusions

The distributed program for RSA enciphering and deciphering achieves nearly-perfect efficiency while running on as many as forty processors. The other distributed programs are all reasonably efficient when running on as many as ten to twenty processors.

From a practical standpoint, the distributed algorithm for enciphering and deciphering is the most important of our algorithms. For primes of the size currently used, parallel execution of the other applications may only be warranted when parallel hardware, that would otherwise be idle, is readily available.

Currently, standard digital signal processing chips can achieve enciphering rates of more than 19 kilobits per second, for a minimum-security RSA cryptosystem with a 512-bit modulus. Expensive special-purpose RSA chips can currently achieve enciphering rates as high as 50 kilobits per second with a 512-bit modulus [Beth 91].

Much higher enciphering rates could be achieved by using many of these chips to encipher in parallel. At some point in the future, it will likely become economically feasible to build dedicated cryptography boards with five or ten special-purpose chips. The enciphering algorithm described above could efficiently use those chips to implement parallel RSA enciphering. Other basic tasks related to the RSA cryptosystem could also be implemented in parallel.

My success in finding two programming paradigms applicable to problems drawn from a single application area confirms the conclusion of Brinch Hansen [93]: programming paradigms can serve a useful role in the development of distributed programs. One surprising result was that I was able to use *both* paradigms, together, to develop a single application: the generation of deterministically certified primes.

For both the competing servers array and the synchronized servers pipeline, my work fit a single pattern. I first spent a substantial amount of time developing, testing, debugging and optimizing a distributed program for a particular application. With just a little effort, I was able to convert the program into a distributed generic program. Subsequently, I could very quickly modify the generic program to implement a new application. The difficult work required to distribute the problem was encapsulated in the generic program.

This suggests that the study of programming paradigms can provide some very practical benefits. With a suitable library of generic programs, it should be possible to quickly develop distributed programs for many applications.

Chapter 2

The RSA Public-Key Cryptosystem

"Only trust thyself, and another shall not betray thee."

William Penn

2.1 Public-Key Cryptography

In 1976, Diffie and Hellman introduced the idea of a public-key cryptosystem [Diffie 76]. Subsequently, Rivest, Shamir and Adleman proposed the first example of a public-key cryptosystem, now known as the RSA cryptosystem [Rivest 78b].

Public-key cryptosystems are based on *trap-door one-way* functions. A *one-way* function is a function for which forward computation is easy, while the backward computation is very hard. In other words, a function $f: X \to Y$ (where X and Y are arbitrary sets) is one-way if it is easy to compute $f(x)$ for every $x \in X$, while it is hard for most $y \in Y$ to find any $x \in X$ such that $f(x) = y$ [Brassard 88].

A function is said to be trap-door one-way if it is possible to easily perform both forward and backward computation, yet the algorithm for backward computation cannot be easily determined without certain secret information, even with knowledge of the complete algorithm for forward computation. In a public-key cryptosystem, an individual makes public an algorithm for the forward computation, allowing anyone to encipher a message using the algorithm. The algorithm for backward computation is kept secret, so that only an authorized individual is capable of deciphering a message enciphered by the public algorithm [Brassard 88].

It has not been proven that any one-way functions exist, but some known functions are candidates. For example, integer multiplication is very easy, while the inverse function, integer factoring, is currently considered to be very hard. The existence of one-way functions is related to the $P = NP$ question [Brassard 88].

2.2 Overview of the RSA Cryptosystem

The RSA cryptosystem uses modular exponentiation with fixed exponent and modulus for enciphering and deciphering. The security of the cryptosystem is based on the belief that such modular exponentiation constitutes a trap-door one-way function [Brassard 88].

Three natural numbers $\langle e, d, M \rangle$ define a particular instance of an RSA cryptosystem, where e is the public enciphering exponent, d is the secret deciphering exponent, and M is the modulus. A plaintext message m (assumed to be in the form of an integer that is greater than 1 and less than M) is enciphered into a cryptogram $c = E(m)$, where $E(m) = m^e \bmod M$. The cryptogram may subsequently be deciphered to retrieve the plaintext message $m = D(c)$, where $D(c) = c^d \bmod M$. $\langle e, M \rangle$ is called the *public-key* and $\langle d, M \rangle$ is called the *secret-key*.

The modulus M is chosen to be the product of two large primes, p and q. As we shall see in the next section, this allows a cryptographer to publish an RSA public-key, without revealing the secret-key.

2.3 How the RSA Cryptosystem Works

The RSA cryptosystem exploits a property of modular arithmetic described by *Euler's Generalization of Fermat's Theorem*. Fermat made the following conjecture, now known as *Fermat's Theorem* [Niven 80]:

Fermat's Theorem: If p is prime, and $\gcd(a, p) = 1$ then $a^{p-1} \equiv 1 \pmod{p}$.

We use $\gcd(a, p)$ to mean the *greatest common divisor* of a and p. When $\gcd(a, p) = 1$ we say that a is *relatively prime* to p. Euler generalized Fermat's theorem into a form that applied to both prime and non-prime moduli [Niven 80]:

Euler's Generalization: If $\gcd(a, n) = 1$ then $a^{\phi(n)} \equiv 1 \pmod{n}$.

Here, $\phi(n)$ is known as *Euler's totient function*, and is defined to be the number of non-negative integers less than n that are relatively prime to n [Niven 80]. For a prime p, every integer from 1 to $p-1$ is relatively prime to p. Accordingly, $\phi(p)$ is defined to be $p-1$.

For an RSA modulus $M = pq$, $\phi(M)$ is easily computed from p and q. There are $M-1$ positive integers less than M. Of those integers, $p-1$ are divisible by q:

$$q, \ 2q, \ 3q, \ ..., \ (p-1)q$$

and $q-1$ are divisible by p:

$$p, \ 2p, \ 3p, \ ..., \ (q-1)p$$

Therefore, we have

$$\phi(M) = (M-1) - (p-1) - (q-1)$$
$$= pq - p - q + 1$$
$$= (p-1)(q-1)$$

Now, recall that using the RSA cryptosystem, a message m is enciphered into a cryptogram

$$c = m^e \bmod M \qquad (2.1)$$

and similarly, the cryptogram is deciphered to reveal the message

$$m = c^d \bmod M \qquad (2.2)$$

Substituting (2.1) into (2.2) gives us

$$m = c^d \bmod M$$
$$= \left(m^e \bmod M\right)^d \bmod M$$
$$= m^{ed} \bmod M$$

This specifies a relationship between e, d, and M necessary for the cryptosystem to function. The keys must be chosen so that

$$m = m^{ed} \bmod M \qquad (2.3)$$

If the two prime factors of M have approximately 100 digits each, the number of possible messages that can be enciphered is approximately 10^{200}. Since $M = pq$, $(p-1) + (q-1) \approx 10^{100}$ integers between 1 and M are not relatively prime to M. So the probability that any given message shares a factor with M is approximately $10^{100}/10^{200} = 10^{-100}$. Consequently, even if an RSA modulus were used to encipher a million different messages, the probability that any of the messages would share a factor with the modulus would be negligibly small.

Therefore, we may reasonably assume that an arbitrary message m is relatively prime to modulus M. By Euler's Generalization, we have

$$m^{\phi(M)} \equiv 1 \ (\bmod \ M)$$

Now we can show that requirement (2.3) is satisfied when the keys are chosen so that

$$ed \equiv 1 \ (\bmod \ \phi(M))$$

which is equivalent to

$$ed = u\phi(M) + 1 \tag{2.4}$$

for some positive integer u.

When (2.4) is true, we have

$$
\begin{aligned}
m^{ed} \bmod M &= m^{u\phi(M)+1} \bmod M && \textit{by (2.4)} \\
&= \left(m^{u\phi(M)} \bmod M\right)\left(m^1 \bmod M\right) \\
&= \left(m^{u\phi(M)} \bmod M\right)m && \textit{by } 1 < m < M \\
&= \left(\left(m^{\phi(M)} \bmod M\right)^u \bmod M\right)m \\
&= \left(1^u \bmod M\right)m && \textit{by Euler's Generalization} \\
&= m
\end{aligned}
$$

and requirement (2.3) is satisfied.

A set of RSA keys is created by first constructing the modulus as the product of two primes ($M = pq$). $\phi(M)$ is easily computed from the prime factors p and q, as described above.

Next, an enciphering exponent e, relatively prime to $\phi(M)$, is chosen. Finally, the deciphering exponent d is computed so that $ed \equiv 1 \pmod{\phi(M)}$. Such a d is called the *inverse* of e modulo $\phi(M)$. When $\phi(M)$ and e are known, this inverse can be found by means of a fast extended gcd algorithm [Denning 82, Knuth 82].

There is no known method for computing $\phi(M)$ without knowing the factors of M. Furthermore, there is no known algorithm to compute an inverse modulo $\phi(M)$ without $\phi(M)$ [Bressoud 89].

RSA security requires, therefore, that modulus M be chosen so that it cannot be factored. If a cryptanalyst could factor the modulus, $\phi(M)$ and d could be easily computed, and the system would be broken.

To thwart a factoring attack, the modulus is chosen to be the product of two large primes. By making the prime factors sufficiently large, general-purpose factoring algorithms are rendered useless [Bressoud 89, Riesel 85, Silverman 91]. In addition, the prime factors are chosen to have special properties that make the modulus safe from special-purpose factoring attacks [Gordon 84, Rivest 78a, Chapter 5].

Given current technology, the prime factors are, typically, chosen to have in the range of seventy-five to one hundred and fifty decimal digits, each [Beth 91, Brassard 88]. The choice of size represents a balance between security from factoring attacks and speed of enciphering and deciphering. The execution times for general-purpose factoring algorithms are exponential in the number of digits of the modulus [Bressoud 89, Riesel 85, Silverman 91]. On the other hand, the running time for RSA enciphering and deciphering, using a standard modular exponentiation algorithm, is cubic [Bressoud 89].

As a result, the time required to factor a modulus grows much faster with modulus size than does the time required to perform enciphering and deciphering operations. Nonetheless, the cubic increase in the time required to perform modular exponentiation is large enough to create practical concerns over the size of the modulus [Beth 91, Brassard 88]. For example, on a 25 MHz T800 transputer [INMOS 88a], modular exponentiation requires just over ten seconds for a modulus and exponent of 640-bits (just under two hundred decimal digits). This corresponds to a sequential enciphering and deciphering rate of just 64 bits per second [Chapter 8].

2.4 Authenticity

Due to the symmetry of the enciphering and deciphering algorithms, the RSA cryptosystem may be used for *authenticity* in addition to secrecy [Denning 82]. Authenticity applications require that a receiver of a message be able to verify that an authorized sender produced the message. In other words, the receiver must be able to determine that a message is authentic, and not a forgery. It is not necessarily important that a message be kept secret. For example, banking transactions may not require secrecy, but undoubtedly require that only an authorized individual be able to initiate a transaction.

For authentication using the RSA cryptosystem, the secret-key can be used to encipher a message. Anybody with the matching public-key will then be able to decipher the message (so it will not be secret). The fact that the public-key deciphers the cryptogram into a meaningful message implies that the cryptogram was produced using the secret-key. In turn, this implies that the message was produced by an authorized individual.

If both secrecy and authenticity are required for a particular application, then two separate sets of RSA keys will be required. One set can be used to encipher a message so that its contents are secret, and the other set can be used to encipher the resulting cryptogram so that the cryptogram's authenticity can be verified.

Chapter 3

Notation for Distributed Algorithms

"Speak clearly, if you speak at all; carve every word before you let it fall."

Oliver Wendell Holmes

3.1 Introduction

To present our algorithms, we use a notation that combines Pascal [Wirth 71] and variants of the distributed programming notations used by Joyce [Brinch Hansen 87, 89] and occam 2 [INMOS 88b]. The notation includes statements for process creation and termination, synchronous communication, and non-deterministic polling. This chapter summarizes the notation used.

3.2 Process Creation and Termination

We use the statement,

  parbegin S1|S2|S3|...|Sk **end**

to denote the execution of k component statements $S1$, $S2$, $S3$, ..., Sk as parallel processes. Execution of a *parbegin* statement terminates only after execution of each component statement has terminated.

 The statement,

  parfor i:=1 **to** k **do** S(i)

is equivalent to

  parbegin S(1)|S(2)|S(3)|...|S(k) **end**

3.3 Process Communication

Parallel processes do not share variables but may share synchronous channels through which they communicate by transferring messages, called *symbols*. A symbol consists of a symbol name along with a (possibly empty) list of data values.

A channel's alphabet is the set of all symbols which the channel may transfer. Similarly to Joyce, a type definition

```
T = [s1(T1), s2(T2), s3(T3), ..., sn(Tn)];
```

defines the alphabet of channel type T. A channel of type T may transfer symbols named $s1$, $s2$, $s3$, ..., sn, carrying values of types $T1$, $T2$, $T3$, ..., Tn, respectively. $T1$, $T2$, $T3$, ..., Tn, are each a (possibly empty) list of type names separated by commas. If a symbol's list of types is empty, parentheses are omitted from the symbol declaration. Such a symbol may be used as a synchronization signal.

For example, a channel of type

```
T = [data(integer,char), eos];
```

may transfer a symbol named *data*, carrying an integer value and a character value. It may also transfer a symbol named *eos*, carrying no value.

Two processes communicate when one process is ready to execute an input statement

```
inp?si(x)
```

and another process is ready to execute an output statement

```
out!si(e)
```

where, *inp* and *out* denote the same channel with type T, the variable x and the expression e are of the same type Ti, and si is the name of a symbol carrying a value of type Ti and in the alphabet of channel type T. The operations denoted by two such communication statements are said to *match*. The communication consists of assigning the value of e to x.

For the communication of a symbol carrying more than one value, an operation corresponding to an input statement

```
inp?si(x1, x2, x3, ..., xk)
```

matches the operation corresponding to an output statement

```
out!si(e1, e2, e3, ..., ek)
```

when *inp* and *out* denote the same channel with type *T*, and the variable list *x1*, *x2*, *x3*, ..., *xk* and expression list *e1, e2, e3, ..., ek* match the same type list *Ti*, associated with symbol name *si*, in the alphabet of channel type *T*.

Variable list *x1, x2, x3, ..., xk* matches type list *t1, t2, t3, ..., tk* when the two lists have the same number of elements, *k*, and for all *i* from 1 to *k*, variable *xi* is of type *ti*. Similarly, expression list *e1, e2, e3, ..., ek* matches type list *t1, t2, t3, ..., tk* when the two lists have the same number of elements, *k*, and for all *i* from 1 to *k*, expression *ei* has a value of type *ti*.

In this case, the communication consists of assigning the value of each expression in the expression list to the corresponding variable in the variable list.

For the communication of a symbol carrying no value, an operation corresponding to an input statement

```
inp?si
```

matches the operation corresponding to an output statement

```
out!si
```

when *inp* and *out* denote the same channel with type *T*, and *si* is the name of a symbol carrying no data, in the alphabet of channel type *T*. The communication consists solely of delaying the two processes executing the operations until execution of the input and output operations may be completed simultaneously.

3.4 Polling

Similarly to Joyce, non-determinism in communication is expressed by a *polling statement*

```
poll G1->S1|G2->S2|G3->S3|...|Gk->Sk end
```

where *S1, S2, S3, ..., Sk* are statement lists, and *G1, G2, G3, ..., Gk* are called *guards*. Each guard is either a *communication guard* or a *timeout guard*.

A communication guard is a communication statement (that is, an input statement or an output statement). During execution of a polling statement, a communication guard is *ready* when some other process is ready to perform a matching, non-polling communication operation.

A timeout guard has the form of

```
after delay
```

During execution, a timeout guard is *ready* when at least *delay* time units have passed since execution of the polling statement began. (Time is measured in arbitrary units called ticks.)

Execution of a polling statement is delayed until at least one of its guards is ready, at which time, a ready guard Gi is selected. If two or more guards are ready, one of the ready guards is arbitrarily selected.

If the selected guard is a communication guard, the corresponding communication is performed, and the guard's associated statement list Si is executed. If the selected guard is a timeout guard, the associated statement list Si is executed.

Chapter 4

The Competing Servers Array

> *"When I was young I observed that nine out of every ten things I did were failures, so I did ten times more work."*
>
> George Bernard Shaw

4.1 Introduction

This chapter describes a programming paradigm that we call the *competing servers* paradigm. A generic program for the paradigm uses multiple server processes that compete against one another to find a solution to a problem and return the solution to a master process. The generic program is applicable to problems which can be implemented as a number of independent searches, and from which only one result is required. The program uses non-determinism to improve parallel efficiency.

We shall use our program shell for the competing servers paradigm to produce candidates for primality testing and to test RSA moduli for susceptibility to special-purpose factoring algorithms.

4.2 Searches Suitable for Competing Servers

As we stated above, the competing servers paradigm includes problems which can be implemented as a number of independent searches, and from which only one result is required. However, the problem need not be a search in the conventional sense. For example, we shall use competing servers to implement a distributed factoring algorithm. Though factoring may not normally be described as a search, for our purposes, it is satisfactory to consider the process of factoring to be a search for factors of a given product.

Other problems suitable to a competing servers implementation include automatic test pattern generation for fault detection in combinational circuits [Ali 91, Hartmann 91] and symbolic processing applications such as theorem proving and expert systems [Natarajan 89].

In general, when competing servers are used to implement a search, more than one server process may find a solution. (These solutions may or may not be distinct, depending upon the nature of the particular application.) As we have specified the problem, exactly one solution is returned to the master. However, the program shell can be varied to allow the return of more than one solution. The program shell can also be varied to handle problems where no solution exists or to allow termination of a computation that has failed to produce a solution after a specified period of time. We shall use the latter variation for our application of competing servers to factoring.

4.3 Overview of the Competing Servers Array

A distributed implementation of the competing servers paradigm uses a number of server processes, along with a master process that coordinates execution of the servers. Each server process must be able to communicate with the master process but does not need to communicate directly with any other server process. Of course, the master process must be able to communicate with each of the server processes.

There are many ways to implement connections that allow the necessary communications; however, we shall limit our choices to implementations that reasonably restrict the number of direct connections between processes. In general, we assume it will not be possible to connect the master process directly to an arbitrary number of server processes. This assumption reflects the current reality for multicomputers, such as the Meiko Computing Surface [Meiko 88] with T800 transputers [INMOS 88a] that we shall use to implement our distributed programs. The Computing Surface allows any processor to be directly connected (via a hardware communication channel) to at most four other processors.

Of the available choices for implementing the competing servers, we choose to connect them as a linear array. We choose a linear array because the communication time is negligible compared to the computing time (at least for the problems of prime generation and factoring) and because the linear array will be convenient for our subsequent implementation of strong prime generation on the Meiko Computing Surface. Figure 4.1 shows a linear array of servers numbered from 1 to p.

Since the master process is not directly connected to every server process, the server processes will have to perform message routing, in order to implement the necessary logical connections. To clearly separate routing from the local search, we break the server process up into several subprocesses.

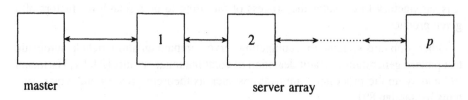

Figure 4.1: Master and server array for the competing servers paradigm

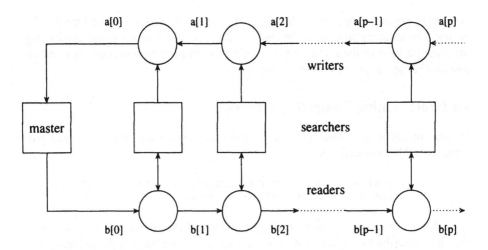

Figure 4.2: *Detailed structure of the competing servers array*

Each server process is broken into three parallel processes: a *reader* process that routes input, a *writer* process that routes output, and a *searcher* process, as shown in figure 4.2. Each instance of the same kind of process executes the same procedure. A pair of unused channels are connected to the right-most server process in the array.

Figure 4.3 identifies the connections for a single server process in the array. For

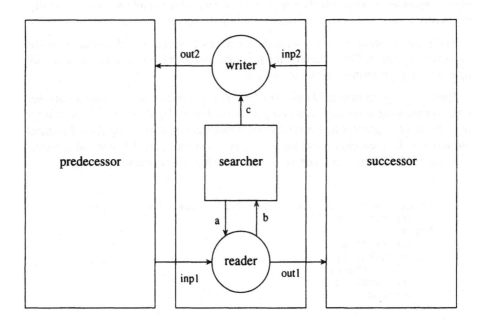

Figure 4.3: *Detailed view of a server process*

any server process in the array, the adjacent process on the left is called the *prede-cessor*. The adjacent process on the right, if any, is called the *successor*. Depending upon the position of the server process within the array, the predecessor may be either another server process, or the master process.

4.4 Implementing Competing Servers

We start by defining a channel protocol for the communicating processes in our program shell. The type definition

```
channel = [start(SearchDataTypes), result(ResultTypes),
    check, continue, complete, eos];
```

is used throughout the program, and is presumed to have been previously declared. We shall describe the meaning of each message as we use them in the processes below.

The *master* process executes algorithm 4.1. Parameter x contains the data necessary to start a search. y is a variable parameter to which the result of a search is assigned. *inp* and *out* are a pair of channels connecting the master to the array of servers.

The master initiates a search by broadcasting search data x as a *start* message. Subsequently, the master awaits a *result* message, containing the single result y returned by the server array. After receiving the result, the master broadcasts a *complete* message, signaling that a result has been found. The master then waits for the server processes to terminate their searches and return the completion signal. Finally, the master broadcasts an *eos* message, signaling the server processes to terminate.

The program shell uses two separate messages to signal search completion and process termination. Consequently, the program shell can be used to implement multiple searches, performed one after another.

Each *server* process consists of a *reader* process, a *writer* process, and a *searcher* process executing in parallel. A server process is defined by algorithm 4.2. Parameter *id* is the server's process number. Channel parameters *inp1*, *out1*, *inp2*, *out2* connect the server to its predecessor and successor, as shown in figure 4.3. Internal channels a, b, and c are created to connect the reader, searcher, and writer to one another.

```
procedure master(x:SearchDataTypes; var y:ResultTypes;
    inp,out:channel);
begin
    out!start(x);
    inp?result(y);
    out!complete;
    inp?complete
    out!eos
end;
```

Algorithm 4.1: Master procedure for competing servers

```
procedure server(id:integer; inp1,out1,inp2,out2:channel);
var a,b,c:channel;
begin
    parbegin
        reader(id, inp1, out1, a, b) |
        searcher(id, b, a, c) |
        writer(id, c, inp2, out2)
    end
end;
```

Algorithm 4.2: The server procedure

Each *reader* process executes algorithm 4.3. The parameters of the reader are the process number *id* and channels *inp1*, *out1*, *inp2*, *out2*. The process number allows each instance of a reader process to determine whether or not it is at the end of the server array. The channels connect the reader to other processes as shown in figure 4.4. *inp1* and *out1* connect the reader to its predecessor and successor, respectively. Therefore, *inp1* acts as the (possibly indirect) connection to the master process. *inp2* and *out2* connect the reader to its local *searcher* process.

The reader repeatedly polls *inp1* for either a *start* message or an *eos* signal. After receiving a *start* message, the reader forwards the message to its successor (if any), and sends the *start* message to the local searcher. The reader then executes a procedure (*monitor*, described below) to interact with the searcher and terminate the search.

After receiving an *eos* signal, the reader passes the signal to its successor (if any) and to the local searcher. It then terminates.

The constant *p* denotes the number of server processes.

During a search, each searcher periodically communicates with its local reader, in order to determine whether a result has been found by some other process. The *moni-*

```
procedure reader(id:integer; inp1,out1,inp2,out2:channel);
var x:SearchDataTypes;
    more:boolean;
begin
    more:=true;
    while more do
        poll
            inp1?start(x) ->
                if id<p then out1!start(x);
                out2!start(x);
                monitor(id, inp1, out1, inp2, out2) |
            inp1?eos ->
                if id<p then out1!eos;
                out2!eos;
                more:=false
        end
    end;
```

Algorithm 4.3: The reader procedure

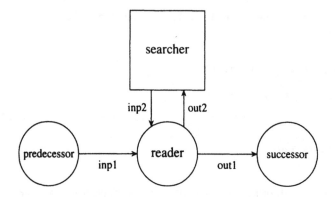

Figure 4.4: *Reader process connections*

tor procedure of algorithm 4.4 defines the reader's interaction with the searcher during a search. The parameters are identical to those of the reader procedure.

During the monitor operation, the reader repeatedly polls its predecessor and its local searcher. If a *check* signal is received from the searcher, a *continue* signal is returned to the searcher. This indicates that a result has not yet been found, and the searcher should continue its search.

If a *complete* signal is received from the searcher, the searcher has successfully completed its search. The reader passes the completion signal to its successor (if any). The reader then awaits a *complete* signal from the master process. Subsequently, the monitor operation terminates.

If a *complete* signal is received from the predecessor, a solution has been returned

```
procedure monitor(id:integer; inp1,out1,inp2,out2:channel);
var more:boolean;
begin
    more :=true;
    while more do
        poll
            inp1?complete ->
                if id<p then out1!complete;
                poll
                    inp2?check -> out2!complete |
                    inp2?complete ->
                end;
                more:=false |
            inp2?check -> out2!continue |
            inp2?complete ->
                if id<p then out1!complete;
                inp1?complete;
                more:=false
        end
end;
```

Algorithm.4.4: *The monitor procedure*

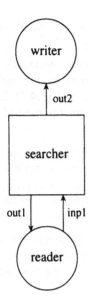

Figure 4.5: *Searcher process connections*

to the master. The reader passes the completion signal on to its successor (if any). The reader then polls the local searcher. If a *check* signal is received, a *complete* signal is returned to terminate the search. If a *complete* signal is received, no message is returned. In this case, the local search process has found a solution, and has already terminated its search. Finally, the monitor operation terminates.

Each *searcher* process is defined by algorithm 4.5. Parameter *id* is the searcher's process number. Channel parameters *inp1*, *out1*, and *out2* connect the searcher to other processes as shown in figure 4.5. *inp1* and *out1* connect the searcher to the local reader process. *out2* connects the searcher to the *writer* process.

A searcher repeatedly polls its local reader for either a *start* message or an *eos* signal. After receiving a *start* message, the searcher executes a procedure (*find*) that

```
procedure searcher(id:integer; inp1,out1,out2:channel);
var x:SearchDataTypes;
    more:boolean;
begin
    more:=true;
    while more do
        poll
            inp1?start(x) ->
                find(id, x, inp1, out1, out2) |
            inp1?eos ->
                out2!eos;
                more:=false
        end
end;
```

Algorithm 4.5: *The searcher procedure*

implements the search to be performed. After receiving an *eos* signal, the searcher passes the signal on to the local writer and then terminates.

The *find* procedure is algorithm 4.6. Parameter *x* is the data necessary to start the search. The remaining parameters contain the process number and the three channels for communication with the local reader and writer processes (Fig. 4.5). The process number can be used to allow each searcher to select a distinct portion of the search space.

The find procedure implements a search that is periodically stopped to check whether it should be terminated. As described above, the periodic check is performed by interacting with the local reader. A *check* signal is sent to the reader. If a *complete* signal is returned, the search is terminated. If a *continue* signal is returned, the search is continued.

If a solution is found, a *result* message containing the solution is sent to the local writer. A *complete* signal is then sent to the reader, and the search is terminated. In any case, execution of the find operation ends with the output of a *complete* signal to the local writer.

initialize, *test* and *update* are application-specific procedures that are defined to implement the search for a given application. *z* represents any additional data that a particular application requires.

The *initialize* procedure defines any necessary initialization to be performed at the start of a search. The *test* procedure implements a step in the search, and returns a result indicating whether or not a solution has been found. The *update* procedure im-

```
procedure find(id:integer; x:SearchDataTypes;
    inp1,out1,out2:channel);
var y:ResultTypes;
    z:InternalDataTypes;
    more,found:boolean;
begin
    initialize(y, z, x, id);
    more:=true;
    while more do begin
        out1!check;
        poll
            inp1?complete -> more:=false |
            inp1?continue ->
                test(y, z, found)
                if found then begin
                    out2!result(y);
                    out1!complete;
                    more:=false
                end;
                update(y, z)
        end
    end;
    out2!complete
end;
```

***Algorithm 4.6**: The find procedure*

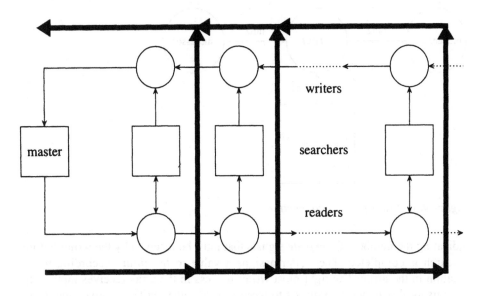

Figure 4.6: Flow of the search completion signal, complete

plements any necessary updating of application-specific data, to be performed after the completion of each step in the search.

The search completion signal is propagated through the reader, searcher and writer processes as shown in figure 4.6.

Each writer process defines a variable *terminate* that acts as a local constant,

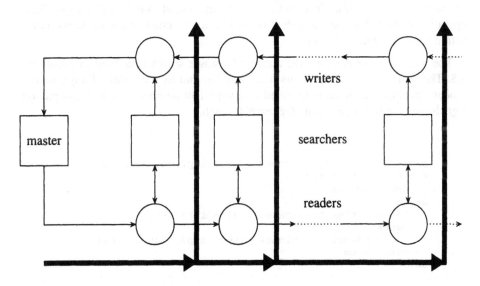

Figure 4.7: Flow of the process termination signal, eos

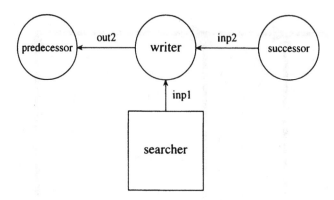

Figure 4.8: *Writer process connections*

indicating the number of *complete* signals that must be received by the writer before a search is completed. The writer assigns a value to *terminate* according to its position in the array. The right-most writer process in the array receives input only from its local searcher. It assigns 1 to *terminate*. All other writers receive input from both a local searcher and from a successor. They assign 2 to *terminate*.

Process termination signal *eos* flows through the processes as shown in figure 4.7. Further details of the two termination signals are discussed later.

Each writer process executes algorithm 4.7. Parameter *id* is the process number. Channel parameters *inp1*, *inp2*, and *out2* connect the writer to other processes as shown in figure 4.8. *inp1* connects the writer to the local searcher. *inp2* and *out2* connect the writer to its successor (if any) and predecessor, respectively.

The writer is designed to handle multiple searches. The main procedure repeatedly executes a procedure (*PassResult*) that handles the output from a single search. That operation updates boolean variable *done* to indicate whether or not a process termination signal has been received.

The procedure for handling the output of a single search is defined by algorithm 4.8. The writer repeatedly polls its local searcher and its successor (if any), until a process termination signal is received or a sufficient number of search completion signals are received to terminate the current search.

```
procedure writer(id:integer; inp1,inp2,out2:channel);
var terminate:integer;
    done:boolean;
begin
    if id<p then terminate:=2 else terminate:=1;
    repeat
        PassResult(terminate, inp1, inp2, out2, done);
    until done
end;
```

Algorithm 4.7: *The writer procedure*

```
procedure PassResult(terminate:integer;
    inp1,inp2,out2:channel; var done:boolean);
var y:ResultTypes;
    count:integer;
    passing:boolean;
begin
    done:=false;
    passing:=true;
    count:=0;
    while not done and (count<terminate) do
        poll
            inp1?result(y) ->
                if passing then begin
                    out2!result(y);
                    passing:=false
                end |
            inp2?result(y) ->
                if passing then begin
                    out2!result(y);
                    passing:=false
                end |
            inp1?complete ->
                count:=count+1;
                if count=terminate then out2!complete |
            inp2?complete ->
                count:=count+1;
                if count=terminate then out2!complete |
            inp1?eos -> done:=true
        end
end;
```

Algorithm 4.8: Processing output from a single search

Since only one result is to be returned to the master process (per search), each writer forwards at most one result. Boolean variable *passing* indicates whether the writer should forward a result to its predecessor. If a first result is received, it is forwarded. If a second result is subsequently received, it is not forwarded.

Finally, the competing servers array is just the parallel composition of the master process and the array of server processes, as defined in algorithm 4.9 (see Fig. 4.2).

```
procedure CompetingServers(x:SearchDataTypes;
    var y:ResultTypes);
type net=array [0..p] of channel;
var a,b:net;
    i:integer;
begin
    parbegin
        master(x, y, a[0], b[0]) |
        parfor i:=1 to p do
            server(i, b[i-1], b[i], a[i], a[i-1])
    end
end;
```

Algorithm 4.9: The competing servers array

4.5 Search Completion and Process Termination

The competing servers array uses two termination signals. The *complete* signal is used to terminate a single search, while the *eos* signal is used to terminate program execution.

When terminating a search, all *result* messages produced by the searchers must be propagated through the writer processes. If this is not implemented carefully, deadlock could result. To avoid deadlock, the search completion signal is forwarded in a manner that insures that all results are output before the completion signal is propagated through the writers (Fig. 4.6).

We can show by induction that all results are output prior to the completion signal. The basis of the induction is simply that a searcher process always outputs its result (if any) before sending a *complete* signal to the local writer process. That this is true is obvious by inspecting algorithm 4.6, which always sends a completion signal to the writer process as its last action.

Now, consider a writer process. The writer receives input from a local searcher, and from another writer. (In the case of the right-most writer, input is received only from a local searcher.) For the moment, assume that the writer receives input only from processes that always output any results prior to the completion signal. Since the writer waits to receive a completion signal over each of its inputs before outputting a completion signal, the writer process must also output any results prior to the completion signal.

We can easily see that each writer does indeed receive input only from processes that output results before the completion signal. The basis step demonstrated that the searcher processes conform to this requirement. By induction, we can also see that the right-most writer receives input only from a searcher. Therefore, its output conforms to the requirement. In turn, the predecessor of the right-most writer also conforms to the output requirement. By induction from right to left, then, each of the writers conforms to the requirement that results are output prior to the completion signal.

The process termination signal, *eos*, is simpler. Since process termination occurs after search completion (or possibly, before any search is started at all), there can be no results remaining to be output. Therefore, we need only insure that each process receives a process termination signal.

When a reader receives a termination signal, it outputs the termination signal to both its local searcher process and to its successor (if any), before terminating. Subsequently, each searcher outputs the termination signal to its local writer process before terminating. Finally, each writer terminates upon receiving the termination signal from its local searcher (Fig 4.7).

4.6 Timing Out

As we mentioned previously, the program shell can be varied to allow a search to be

```
procedure master(x:SearchDataTypes; timeLimit:integer;
   var y:ResultTypes; var found:boolean; inp,out:channel);
begin
   out!start(x);
   found:=true;
   poll
       inp?result(y) ->
           out!complete;
           inp?complete |
       after timeLimit ->
           out!complete;
           poll
               inp?result(y) -> inp?complete |
               inp?complete -> found:=false
           end
   end;
   out!eos
end;
```

Algorithm 4.10: *Modified master procedure that supports timing out*

timed out. That is, if a search has failed to find a solution after a specified period of time, the master can terminate the search. This variation is of particular interest to us, since we use it to develop a distributed factoring program based on competing servers.

The changes necessary for this variation are trivial. The server processes do not have to be modified at all. The time-out variation uses the modified master process defined by algorithm 4.10.

The modified version of the master requires two additional parameters. *timeLimit* is an integer parameter specifying the time limitation for the search. Variable parameter *found* is used to indicate whether or not a solution was found.

The modified master process executes a polling statement with a timeout guard to detect whether the time limit for the search is exceeded. In that case, the master broadcasts a search completion signal to terminate the search. Subsequently, the master polls the input channel for either a result or for the return of the completion signal. (Recall that propagation of the completion signal may flush out a result.) If no result is received, the master updates boolean variable *found* to indicate that no solution was found.

For the time-out version of the competing servers array, algorithm 4.11 replaces algorithm 4.9. All other algorithms remain unchanged.

4.7 Determinism versus Non-Determinism

We have presented a generic program for competing servers that relies heavily on non-determinism. It is possible to implement competing servers as a deterministic program without polling of multiple channels. Unfortunately, deterministic communication restricts the class of problems for which competing servers is efficient.

```
procedure CompetingServers(x:SearchDataTypes;
    timeLimit:integer; var y:ResultTypes; var found:boolean);
type net=array [0..p] of channel;
var a,b:net;
    i:integer;
begin
    parbegin
        master(x, timeLimit, y, found, a[0], b[0]) |
        parfor i:=1 to p do
            server(i, b[i-1], b[i], a[i], a[i-1])
    end
end;
```

Algorithm 4.11: *The modified competing servers array*

A deterministic implementation is efficient only for applications where the execution time for each step in the search is roughly constant. This is because each server process ends up waiting for the slowest server process to complete its search step, before continuing. When each server process has roughly the same amount of work to do, this can be efficient. However, when the execution times for steps in the search vary widely, the deterministic implementation is inefficient.

The non-deterministic approach, on the other hand, is useful both for applications where the execution times for search steps are fairly constant and for those where the execution times of the search steps may vary widely.

In the case of the generating prime candidates, non-determinism is important because some numbers being tested will be discarded very quickly, while other numbers will require the application of a much slower primality test [Chapter 5]. Our non-deterministic implementation allows each server to test numbers for prime candidacy as fast as it can, independent of how fast or slow other server processes proceed. (Slow servers can increase the time required to propagate a search completion signal.)

Performance tests show that strong prime generation using the non-deterministic competing servers implementation is twice as fast (and twice as efficient) on forty processors as a strong prime generation program using a deterministic implementation of competing servers.

4.8 Performance

We reprogrammed the generic program for competing servers using occam 2 on a Meiko Computing Surface with T800 transputers, and used it to implement distributed programs for the generation of (large) prime candidates for a strong prime number generator and for special-purpose factoring of RSA moduli.

Table 4.1 shows the average execution time, speedup and efficiency for a distributed program to generate strong prime numbers described in chapter 5. The figures shown are averages over fifty trials for the generation of 512-bit strong primes. The strong prime generation program uses a distributed prime candidate generation

Processors	Execution Time (s)	Speedup	Efficiency
1	725	1.00	1.00
5	158	4.58	0.92
10	92.7	7.82	0.78
20	57.8	12.5	0.63
40	37.7	19.2	0.48

Table 4.1: *Performance figures for parallel strong prime generation*

subprogram based on competing servers along with a distributed prime certification subprogram based on Brinch Hansen's distributed primality testing algorithm [Brinch Hansen 92b,c]. See chapter 5 for a detailed analysis of the performance of the algorithm.

A distributed implementation of a special-purpose factoring algorithm based on competing servers with the timing out modification is described in chapter 6. That implementation exhibits nearly-perfect speedup running on forty processors.

Chapter 5

A Distributed Algorithm to Find Safe RSA Keys

> *"Man can believe the impossible, but can never believe the improbable."*
>
> Oscar Wilde

5.1 Introduction

This chapter summarizes the requirements for safe RSA keys, and presents an algorithm for generating RSA keys that satisfy those requirements. We then describe a distributed implementation of the algorithm based on the competing servers program shell described in chapter 4 and the distributed primality testing program described in [Brinch Hansen 92b].

5.2 Considerations for RSA Keys

While the largest number reported to have been factored using a general-purpose factoring algorithm had only 116 decimal digits [Silverman 91], much larger numbers (and even many numbers which are the product of two large primes) can be successfully factored using one of a number of special-purpose factoring algorithms [Bressoud 89, Riesel 85]. To avoid using a modulus for RSA enciphering that can be easily factored with one of these special-purpose algorithms, extreme care must be exercised in selecting the two prime factors for the modulus.

In addition to special-purpose factoring algorithms, other types of cryptanalytic attacks on RSA can be successful if an unsafe modulus is chosen [Denning 82, Riesel 85]. These attacks, too, must be considered when selecting the prime factors for a modulus.

We call an RSA modulus that is secure from special-purpose factoring algorithms and cryptanalytic attacks a *safe modulus* (or a *strong modulus*). Similarly, we call a prime number that can be used to construct a safe RSA modulus a *safe prime* (or a *strong prime*).

Given a safe RSA modulus, it is a simple task to select a secure pair of enciphering and deciphering exponents for a particular instance of the RSA cryptosystem [Rivest 78a]. Consequently, we are concerned solely with the problem of selecting a safe modulus.

5.3 Characteristics of Strong Primes

Since the security of the RSA cryptosystem requires that the modulus be impossible to factor (using any feasible amount of computing resources), it is essential that the modulus not be susceptible to factoring by a special-purpose factoring algorithm. We consider four basic factoring attacks that can be applied to an RSA modulus.

Several simple factoring techniques have a running time which is polynomial in the size of the smallest prime factor of the product being factored. One simple technique, *trial division*, factors a product by repeatedly attempting to divide the product by increasing numbers, starting at 2, until a factor is found. Clearly, if p is the smallest factor of the product, then trial division is $O(p)$. (More complex factoring algorithms possess running times that are even faster than $O(p)$.) Therefore, we should choose the two prime factors to be approximately the same size (say, in binary digits).

Another simple factoring technique, developed by Fermat, searches for the factors of a product starting at the square root of the product. The algorithm has a running time that increases with the difference between the two prime factors of the product [Bressoud 89, Riesel 85]. Consequently, we should choose the two prime factors to have a substantial difference between them.

We, therefore, should select a pair of primes of approximately the same order of magnitude, but with a significant difference. For example, if we were selecting a modulus of approximately two hundred decimal digits, we might select one prime to be approximately 0.6×10^{100} and the other to be approximately 0.4×10^{100}, in which case, both the smaller prime and the difference between the primes (0.2×10^{100}) are quite large. Assuming that we can generate large primes, these requirements can be easily fulfilled. We, therefore, concern ourselves with the specific properties that make an individual prime factor strong (from the standpoint of constructing a safe RSA modulus).

Towards this end, we consider two other special-purpose factoring algorithms. Consider a product with prime factor p. The first algorithm, *Pollard's $p-1$ method*, can quickly factor the product if $p-1$ has only small prime factors (say, less than one million) [Pollard 74, Chapter 6]. The second algorithm, *Williams' $p+1$ method*, can similarly factor the product if $p+1$ has only small prime factors [Williams 82]. Consequently, for each prime factor p of the modulus, we should insure that both $p-1$ and $p+1$ contain at least one large prime factor. In other words, there should be some large prime r such that $p-1$ is a multiple of r. Similarly, there should be some large prime s such that $p+1$ is a multiple of s.

Therefore, we should choose p such that

$$p \equiv 1 \pmod{r},$$
$$p \equiv s - 1 \pmod{s}.$$

In addition to special purpose factoring algorithms, we should consider other possible attacks upon the RSA cryptosystem. In particular, we consider an attack by which the encryption algorithm is repeatedly applied to a ciphertext produced by RSA encryption. For a poorly chosen modulus, repeatedly applying the encryption algorithm a small number of times (say, less than a million) may successfully decrypt a very large percentage of the enciphered messages [Denning 82, Riesel 85]. While there will always be some messages susceptible to such decryption, a careful choice of modulus can reduce the number of such messages to a negligible amount. Two different approaches have been suggested to achieve this reduction.

One approach suggests that each prime factor, p, of the modulus should be chosen so that $p - 1 = 2r$, where once again, r is a large prime [Blakley 79]. Clearly, this is a special case of our previous requirement that $p - 1$ should be some multiple of r.

The second approach suggests that each prime factor p should be chosen so that $p - 1$ is some multiple of a large prime r. In addition, the prime r should be chosen such that $r - 1$ is some multiple of a large prime t [Rivest 78a]. That is, we should choose p so that

$$p \equiv 1 \pmod{r},$$
$$r \equiv 1 \pmod{t}.$$

As a practical matter, it is easier to generate a prime p that satisfies these requirements than it is to generate a p so that $p - 1 = 2r$. Furthermore, there is a simple technique that allows us to generate a prime that both satisfies the requirements of the second approach and satisfies the additional requirement that $p + 1$ is a multiple of a large prime [Gordon 84].

We, therefore, define p to be a *strong prime* when it satisfies the conditions:

$$p \equiv 1 \pmod{r},$$
$$p \equiv s - 1 \pmod{s},$$
$$r \equiv 1 \pmod{t}.$$

where p, r, s, and t are large primes.

Example: As a small example of a "strong" prime consider the prime

$$p = 199819$$
$$= 2 \times 3^2 \times 17 \times 653 + 1$$
$$= 2^2 \times 5 \times 97 \times 103 - 1$$

and additionally consider that

$$653 = 2^2 \times 163 + 1,$$

where 653, 103, and 163 are primes.

Therefore,

$$199819 \equiv 1 \pmod{653},$$
$$199819 \equiv -1 \pmod{103}$$
$$\equiv 102 \pmod{103},$$
$$653 \equiv 1 \pmod{163},$$

and p is a "strong" prime with component primes

$$r = 653,$$
$$s = 103,$$
$$t = 163.$$

(We use quotes because 199819 is too small to be used in an RSA modulus.)

End Example.

5.4 Constructing Strong Primes

Gordon [84] provided an efficient algorithm to construct strong primes, as defined above. We describe the technique here.

For clarity in this discussion, we must be able to distinguish between primes known to have certain characteristics from other not know to have those characteristics. We call prime r a *double prime*, when it is known to satisfy the requirement $r \equiv 1 \pmod{t}$ for some large prime t. We call a prime that is not known to possess any special properties a *simple prime*.

Our description of a prime as either simple or double is just a statement of our knowledge about the prime. A prime that we call simple or double may actually satisfy our requirements for a strong prime. Our knowledge of that prime, however, is insufficient to conclude whether or not the prime is strong.

For now, let us assume that we can generate simple primes (or test a number for primality) without difficulty. We start the process of constructing strong prime p by generating simple primes s and t. We then generate double prime r such that $r - 1$ is a multiple of t. This is done by computing numbers of the form $kt + 1$ and testing them for primality. The first number of this form found to be prime will be used as r.

Since we are dealing with primes much larger than 2, all of our primes will be odd. Given two distinct, odd primes r and s, we may construct the strong prime p using Theorem 5.1, based upon a theorem from [Gordon 84].

Theorem 5.1: If r and s are odd primes, then prime p satisfies

$$p \equiv 1 \ (\text{mod } r) \equiv s - 1 \ (\text{mod } s)$$

if p is of the form

$$p = u(r,s) + krs$$

where

$$u(r,s) = (s^{r-1} - r^{s-1}) \ \text{mod } rs$$

and k is an integer.

Proof: To prove this, we need to recall *Fermat's Theorem* [Niven 80], which states:

if p is a prime that does not divide x, then $x^{p-1} \equiv 1 \ (\text{mod } p)$.

Since s and r are distinct primes, they cannot divide one another and we have

$$s^{r-1} \equiv 1 \ (\text{mod } r), \tag{5.1}$$

$$r^{s-1} \equiv 1 \ (\text{mod } s). \tag{5.2}$$

Furthermore,

$$s^{r-1} \equiv 0 \ (\text{mod } s), \tag{5.3}$$

$$r^{s-1} \equiv 0 \ (\text{mod } r), \tag{5.4}$$

$$krs \equiv 0 \ (\text{mod } r) \equiv 0 \ (\text{mod } s). \tag{5.5}$$

Now, by definition,

$$\begin{aligned} p &= u(r,s) + krs \\ &= \left(s^{r-1} - r^{s-1}\right) \text{mod } rs + krs \end{aligned} \tag{5.6}$$

For any integers p, q and r, if $p = q$ then $p \equiv q \ (\text{mod } r)$. Taking p modulo r and applying this to equation (5.6) we get,

$$\begin{aligned} p &\equiv \left(\left(s^{r-1} - r^{s-1}\right) \text{mod } rs + krs\right) (\text{mod } r) \\ &\equiv s^{r-1} - r^{s-1} + krs \ (\text{mod } r) \qquad \text{\textit{since rs is a multiple of r}} \\ &\equiv 1 - 0 + 0 \ (\text{mod } r) \qquad\qquad \text{\textit{by (5.1), (5.4) and (5.5)}} \\ &\equiv 1 \ (\text{mod } r) \end{aligned}$$

Similarly, taking p modulo s, we have

$$p \equiv \left(\left(s^{r-1} - r^{s-1} \right) \bmod rs + krs \right) \; (\bmod \; s)$$

$$\equiv s^{r-1} - r^{s-1} + krs \; (\bmod \; s) \qquad \textit{since rs is a multiple of s}$$

$$\equiv 0 - 1 + 0 \; (\bmod \; s) \qquad \textit{by (5.2), (5.3) and (5.5)}$$

$$\equiv -1 \; (\bmod \; s)$$

$$\equiv s - 1 \; (\bmod \; s)$$

End Proof.

Since r and s are both odd, rs is odd and krs is alternately odd and even, with increasing k. So are numbers of the form $u(r,s) + krs$. Since all of our primes are odd, it is convenient to consider p to be of the form

$$p_0 + 2krs,$$

where

if $u(r,s)$ is odd then $p_0 = u(r,s)$, otherwise $p_0 = u(r,s) + rs$.

This insures that p_0 is odd. Since $2krs$ is even for all k, $p_0 + 2krs$ is always odd, and even numbers are excluded from consideration.

This gives us a simple method for generating strong primes. The method is defined by algorithm 5.1. Parameters *seed1* and *seed2* represent the starting points from which we should start looking for simple primes. The operation *modpower(a, b, c)* computes $a^b \bmod c$.

```
function StrongPrime(seed1,seed2:integer):integer;
var p,r,s,t,u,p0,rs:integer;
    k:integer;
    prime:boolean;
begin
    s:=SimplePrime(seed1);
    t:=SimplePrime(seed2);
    r:=DoublePrime(t);
    rs:=r*s
    u:=(rs+modpower(s, r-1, rs)-modpower(r, s-1, rs)) mod rs;
    if odd(u) then p0:=u else p0:=u+rs;
    k:=0;
    repeat
        p:=p0+2*k*rs;
        certify(p, prime);
        k:=k+1
    until prime;
    StrongPrime:=p
end;
```

Algorithm 5.1: Generating strong primes

In computing u, the rs term is added simply to insure that the overall result will be a positive integer. This is necessary because the *mod* operator in most programming languages is not identical to the mathematical operator *mod*. The mathematical operator has the property that $0 \le (a \bmod b) < b$. For most programming languages, however, $a \bmod b$ is negative when a is negative.

5.5 Certifying Primes

Primality testing is the process of distinguishing primes from *composites* (products of more than one prime). In the previous section, we assumed that we could certify large primes without difficulty, as is necessary to construct a strong prime. In this section, we examine the process of certifying large primes.

Our basic tool for certifying primes is the Miller-Rabin witness test based upon Fermat's theorem and quadratic residues [Miller 76, Rabin 80] to test a number for compositeness. Brinch Hansen [92b] describes the method and develops algorithm 5.2.

The Miller-Rabin test uses two techniques to test a number for compositeness. Foremost, it performs a *Fermat test* which determines whether Fermat's theorem is satisfied by the pair x and p. By Fermat's theorem, if $(x^{p-1} \bmod p) \neq 1$, then p must be a composite number. The basic algorithm simply performs the necessary modular exponentiation using a well-known, fast exponentiation algorithm based on successive squaring [Bressoud 89].

In addition, the Fermat test is supplemented by a so-called *quadratic residue test*. It can be shown that if $y^2 \equiv 1 \pmod{p}$ for some integer y, where $1 < y < p - 1$, then p is composite [Brinch Hansen 92b].

The Miller-Rabin algorithm is a probabilistic test for compositeness. That is, the algorithm can prove that a number is composite, but it cannot prove that a number is prime. If the algorithm fails to show that a number is composite, the number may in

```
function witness(x,p:integer):boolean;
var e,m,p1,r,y:integer;
    sure:boolean;
begin {1<=x<p}
    m:=1; y:=x; e:=p-1;
    p1:=e; sure:=false;
    while not sure and (e>0) do
        if odd(e) then
            begin m:=(m*y) mod p; e:=e-1 end
        else begin
            r:=y;
            y:=sqr(y) mod p; e:=e div 2;
            if y=1 then sure:=(1<r) and (r<p1)
        end;
    witness:=sure or (m<>1)
end;
```

Algorithm 5.2: The Miller-Rabin witness test

```
function prime(p:integer;m:integer):boolean;
var sure:boolean;
    i:integer;
begin
    sure:=false;
    for i:=1 to m do
        if witness(random(1, p-1), p) then sure:=true;
    prime:=not sure
end;
```

Algorithm 5.3: *Prime certification*

fact be either prime or composite. However, it turns out that the algorithm almost always identifies a composite as such.

Therefore, if the algorithm does not identify p as a composite, p is almost always a prime. While Rabin has formally proven that 0.25 is a bound on the probability that a composite number will go undetected by the witness algorithm [Rabin 80], this bound is not tight. In practice, it has been observed that the chance that a composite number goes undetected by the algorithm is very close to zero [Beauchemin 86, Gordon 84, Jung 87, Rivest 90].

The generally accepted technique for using the Miller-Rabin test to certify a prime is to repeat the test some arbitrary number of times (m) to make the probability of an error negligible [Beauchemin 86, Gordon 84, Jung 87]. We select $m = 40$, consistent with [Brinch Hansen 92b]. This value of m insures that there will be no chance, in practice, of the Miller-Rabin test failing.

Using Rabin's bound, the probability that 40 Miller-Rabin tests fail to detect a composite as such is conventionally estimated as less than $0.25^{40} \approx 10^{-24}$. For primes of the size we are using, this bound is almost certainly accurate. However, the formal bound on the probability of an error is a bit tricky to compute and depends upon the size of the number being tested for primality [Beauchemin 86].

[Brinch Hansen 92b] develops algorithm 5.3 as the certification algorithm. The *for* statement shows that the m witness tests can be performed simultaneously by m processors. The operation *random(1, p–1)* randomly selects an integer from the range 1 to $p-1$.

5.6 Generating Primes

We are now prepared to examine the process of generating prime numbers. Since Miller-Rabin certification of prime numbers is a lengthy process (requiring repeated use of the fairly slow witness algorithm), we cannot simply apply algorithm 5.3 to various numbers until we happen upon a prime. Rather, we must look for a technique that has two important properties:

    1) it selects numbers in a manner that will, quickly and systematically, lead to the eventual selection of a prime number for certification, and

2) it only attempts to certify those numbers which are almost definitely prime.

The first property is reasonably easy to satisfy. On the average, one in every $\ln n$ numbers close to n is prime [Bressoud 89, Riesel 85]. If we simply select a random n and start counting upward, we will eventually reach a prime. This will take an average of $O(\ln n)$ steps.

The second property is also reasonably easy to satisfy. Recall that, in practice, a single Miller-Rabin witness test almost always detects composite numbers correctly. Therefore, we may simply use a single test to determine whether full certification of a number is worthwhile. This technique will, very reliably, avoid wasted use of the lengthy certification procedure.

A single Miller-Rabin test is, itself, relatively slow, however. Therefore, it is also desirable to use another, faster technique to select those numbers to which the single Miller-Rabin test should be applied. For example, rather than selecting a random number and incrementing by 1, we should start with an odd random number and increment by 2. This decreases the number of single witness tests required by a factor of two.

The logical extension of this is to sieve out any numbers that are multiples of other small primes, 3, 5, 7, ... The simplest way to do this is to perform trial division by small primes. If the selected number can be evenly divided by any of the small primes, it is not prime. Since trial division by small primes is much faster than the modular exponentiation of the witness test, this reduces the computing time substantially.

Algorithm 5.4 shows the certification procedure used for prime generation. Integer p is tested for primality by first testing for small prime factors. If none are found, then a single Miller-Rabin test is used. If the test does not show the number to be composite, then the number is almost surely a prime, and we call it a *prime candidate*. Finally, if the given number is found to be a prime candidate, then a full certification procedure, consisting of forty Miller-Rabin tests, is performed.

Algorithm 5.5 implements the trial division by small primes, where *NumPrimes* is a constant specifying the number of small primes to be used, and *primes* is an array containing the small primes.

Algorithm 5.6 defines an algorithm to generate simple prime y. The algorithm searches for a prime near odd x by first testing x for primality and (assuming that x is not prime) subsequently testing increasingly large odd numbers until a prime is found.

```
procedure certify(p:integer; var passed:boolean);
begin
   passed:=false;
   if not SmallFactor(p) then
      if prime(p, 1) then passed:=prime(p, 40)
end;
```

Algorithm 5.4: Certification procedure for prime generation

```
function SmallFactor(p:integer):boolean;
var j:integer;
    found:boolean;
begin
    found:=false;
    j:=0;
    while not found and (j < NumPrimes) do begin
        found:=(p mod primes[j] = 0);
        j:=j+1
    end;
    SmallFactor:=found
end;
```

Algorithm 5.5: Testing for small factors

```
function SimplePrime(x:integer):integer;
var prime:boolean;
    y:integer;
    k:integer;
begin {odd(x)}
    prime:=false;
    k:=0;
    repeat
        y:=x+k;
        certify(y, prime);
        k:=k+2
    until prime;
    SimplePrime:=y
end;
```

Algorithm 5.6: Simple prime generation

We can use algorithm 5.7 to generate double prime y in a very similar manner. In algorithm 5.7, however, we test increasing odd numbers of the form $y = kx + 1$, which insures that $y - 1$ will be a multiple of x.

5.7 Parallel Algorithm

We shall develop a distributed algorithm by separating the process of generating prime numbers into two problems: the problem of generating prime candidates, and the problem of certifying prime candidates as prime.

A distributed algorithm to solve the latter problem was described in [Brinch Hansen 92b,c]. We develop a distributed algorithm to solve the former problem using the program shell for competing servers [Chapter 4]. Combining the two algorithms properly yields an efficient distributed algorithm to generate strong primes.

We first use competing servers to implement a distributed algorithm to generate simple prime candidates. We then combine that algorithm with the prime certification algorithm, so that it generates certified primes. Finally, we show how to modify the algorithm to generate strong primes.

```
function DoublePrime(x:integer):integer;
var prime:boolean;
    y:integer;
    k:integer;
begin {x is prime}
    prime:=false;
    k:=2;
    repeat
        y:=k*x+1;
        certify(y, prime)
        k:=k+2
    until prime;
    DoublePrime:=y
end;
```

Algorithm 5.7: Double prime generation

5.8 Parallel Candidate Generation

The generic program for competing servers uses a channel protocol defined by the type declaration

```
channel = [start(SearchDataTypes), result(ResultTypes),
    check, continue, complete, eos];
```

When implementing a particular application, we must substitute for *Search-DataTypes* and *ResultTypes*.

The heart of the competing servers program is a sequential search executed by each of the server processes. The search is periodically stopped to check whether the search should be terminated, as a result of some other server having found a solution. Algorithm 5.8 is a modified version of the generic procedure, specialized for use in generating a (simple) prime candidate.

For the process of generating prime candidate y, we start with seed value x, indicating where the search for prime candidates should begin. x is presumed to be an odd integer. Parameter *id* is a process number. The servers are numbered from 1 to p, where p is a constant.

The servers must examine disjoint portions of the search space. To achieve this, each server starts searching at $x + 2(id - 1)$ using increments of $2p$. This interleaves the numbers examined by each server, and allows the server array, as a whole, to examine odd numbers greater than or equal to x (until a prime candidate is found).

The parameters *inp1* and *out1* are channels connecting the searcher process to a local *reader* process. During the search for factors, each searcher periodically communicates with its local reader. The searcher sends a *check* signal to the reader. The reader responds with either a *complete* signal or a *continue* signal. A *complete* signal indicates that the search should be terminated. A *continue* signal indicates that the search should be continued [Chapter 4].

```
procedure FindSimple(id:integer; x:integer;
     inp1,out1,out2:channel);
  var y,k:integer;
      more,found:boolean;
  begin
     k:=(id-1)*2;
     more:=true;
     while more do begin
        out1!check;
        poll
           inp1?complete -> more:=false |
           inp1?continue ->
              y:=x+k;
              found:=false;
              if not SmallFactor(y) then found:=prime(y, 1);
              if found then begin
                 out2!result(y);
                 out1!complete;
                 more:=false
              end;
              k:=k+p*2
        end
     end;
     out2!complete
  end;
```

Algorithm 5.8: Modified find procedure to generate a simple prime candidate

out2 is a channel connecting the *searcher* process to a local *writer* process. If the searcher finds a factor, it outputs that factor to its local writer.

Once a candidate is found, it is returned to the master. We substitute the type *integer* for both *SearchDataTypes* and *ResultTypes*. Together with the remainder of the competing servers program shell described in chapter 4, this defines a complete distributed algorithm to generate prime candidates.

5.9 Parallel Prime Certification

A distributed algorithm for primality testing is developed in [Brinch Hansen 92b] based on the Monte Carlo paradigm described in [Brinch Hansen 92c]. We summarize the algorithm here for completeness.

The Monte Carlo paradigm uses a master process that communicates with a fixed number of server processes. For implementation on a multicomputer, this may be done by creating a pipeline, as shown in figure 5.1.

The master process simply broadcasts a problem, and then collects the results of the Monte Carlo trials performed by the server processes. In the case of prime certification, the master process can simply execute:

```
out!certify(x);
for i:=1 to m do inp?trial(b[i])
```

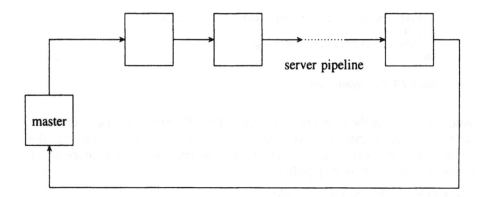

Figure 5.1: *Master and server pipeline for Monte Carlo paradigm (primality testing)*

where *x* is a prime candidate to be certified, *inp* and *out* are channels of type

```
channel = [certify(integer), trial(boolean)];
```

connecting the master process to the server pipeline, *m* is a constant denoting the number of Monte Carlo trials to be performed, and *b* is an array of booleans used to hold the returned results.

Correspondingly, each server process in the primality testing pipeline executes algorithm 5.9. The operation *initialize(id*id)* initializes the random number generator seed. *p* is again a constant denoting the number of server processes on the pipeline. *id* is a process number between 1 and *p*.

A server inputs the number to be certified and outputs the same number to the next server process on the pipeline (if any). The server then performs *q* Miller-Rabin

```
procedure certifier(id:integer; inp,out:channel);
var x:integer;
    b:boolean;
    j,k,q:integer;
begin
    initialize(id*id);
    inp?certify(x);
    if id<p then out!certify(x);
    q:=m div p;
    for j:=1 to q do begin
        solve(x, b);
        out!trial(b);
        for k:=1 to id-1 do begin
            inp?trial(b);
            out!trial(b)
        end
    end
end;
```

Algorithm 5.9: Primality testing server algorithm

```
procedure solve(p:integer; var sure:boolean);
begin
    sure:=witness(random(1, p-1), p)
end;
```

Algorithm 5.10: Procedure solve

witness tests, where the total number of tests to be performed is $m = pq$. (m must be divisible by the number of server processes, p.) After each Miller-Rabin test, the server outputs its local result. The server then transfers the results from the $id - 1$ servers that precede it on the pipeline.

Algorithm 5.10 is the *solve* procedure.

5.10 Combining Generation and Certification

To generate prime numbers, we must both generate prime candidates and subsequently certify those candidates as primes. Consequently, we will need to combine our distributed programs for candidate generation and primality testing. Figure 5.2 shows a single master process with two sets of servers: one for the generation of prime candidates (an array of competing servers), and the other for certify primes (a pipeline of Monte Carlo servers).

For convenience, we combine the two channel protocols into a single protocol defined by the type declaration

```
channel = [start(integer), result(integer), check,
    continue, complete, eos, certify(integer),
    trial(boolean)];
```

Since generating and certifying a prime requires that the master first use the candidate generating servers to produce a prime candidate, and then certify the candidate using the primality testing servers, the master process executes algorithm 5.11.

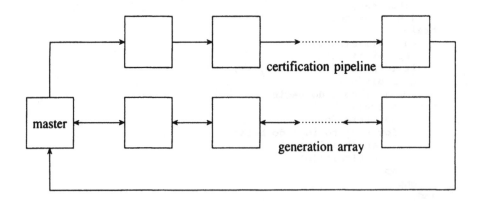

Figure 5.2: Master with two sets of servers

```
procedure master(seed:integer; var p:integer;
   var prime:boolean; inp1,out1,inp2,out2:channel);
type table=array [1..m] of boolean;
var i:integer;
   b:table;
begin
   out2!start(seed);
   inp2?result(p);
   out2!complete;
   inp2?complete;
   out2!eos;
   out1!certify(p);
   for i:=1 to m do inp1?trial(b[i]);
   test(b, prime)
end;
```

Algorithm 5.11: Master procedure for prime generation

inp1 and *out1* are channels connecting the master to the primality testing pipeline. *inp2* and *out2* connect the master to the candidate generating array. The operation *test(b, prime)* sets boolean variable *prime* to indicate whether or not each witness test was passed.

The parallel network combining the master process with the two sets of servers is defined by algorithm 5.12. (The *generator* statement corresponds to the *server* statement in chapter 4.)

To generate a prime, we first generate prime candidate p via a parallel search, and then certify that candidate by performing m Miller-Rabin tests in parallel. Since prime candidate generation and certification do not take place simultaneously, a significant amount of processing power would be wasted if we were to place each server process on its own processor. Since only one of the sets of servers is working at any one time, we can achieve much better efficiency by placing one candidate generating server process and one certification server process on each processor.

```
procedure GeneratePrime(seed:integer; var x:integer;
   var prime:boolean);
type net=array [0..p] of channel;
var a,b,c:net;
   i:integer;
begin
   parbegin
      master(seed, x, prime, a[p], a[0], b[0], c[0])|
      parfor i:=1 to p do
         parbegin
            certifier(i, a[i-1], a[i])|
            generator(i, c[i-1], c[i], b[i], b[i-1])
         end
   end
end;
```

Algorithm 5.12: Arranging the overall computation

To achieve this improvement, we need only insure that the innermost *parbegin* of algorithm 5.12 is executed by a single processor. We also improve efficiency by executing the master and the collector on a single processor.

5.11 Generating Strong Primes

To generate a strong prime, we will need to generate and certify several prime candidates. This may be done by modifying our two sets of servers so that they will solve a sequence of problems. The competing servers array was designed to support multiple searches [Chapter 4], but we do need to create the specialized search algorithms for double prime and strong prime candidate generation.

To accommodate the generation of simple, double and strong prime candidates, the channel protocol is modified to be

```
channel = [simple(integer), double(integer,integer),
    strong(integer,integer,integer), result(integer), check,
    continue, complete, eos, certify(integer),
    trial(boolean)];
```

where the messages *simple*, *double* and *strong* are used to start searches for simple, double and strong prime candidates, respectively.

Algorithm 5.13 shows the procedure of the master process modified for the generation of strong primes. After a strong prime has been generated, the master terminates each set of servers with an *eos* signal.

Algorithm 5.14 is the master procedure to generate a simple prime. A search for a simple prime is initiated with the output of a *simple* symbol to the candidate generating array. Once prime candidate p is received, it is output to the primality testing pipeline. If p is prime, the operation is complete. Otherwise, the process is repeated, with a new search starting at $p + 2$.

Algorithm 5.15 is the master procedure to generate a double prime. It is identical to algorithm 5.14, except that a search is initiated for double prime $r = kt + 1$, for some $k \geq 0$. If received candidate r is not prime, the process is repeated with a new search starting at $k = (r \text{ div } t) + 2$.

```
procedure master(seed1,seed2:integer; var p:integer;
    inp1,out1,inp2,out2:channel);
var r, s, t:integer;
begin
    GenerateSimplePrime(seed1, s, inp1, out1, inp2, out2);
    GenerateSimplePrime(seed2, t, inp1, out1, inp2, out2);
    GenerateDoublePrime(t, r, inp1, out1, inp2, out2);
    GenerateStrongPrime(r, s, p, inp1, out1, inp2, out2);
    out1!eos;
    out2!eos
end;
```

Algorithm 5.13: Generating strong primes

```
procedure GenerateSimplePrime(seed:integer; var p:integer;
   inp1,out1,inp2,out2:channel);
type table=array [1..m] of boolean;
var i:integer;
   b:table;
   prime:boolean;
begin
   repeat
      out2!simple(seed);
      inp2?result(p);
      out2!complete;
      out1!certify(p);
      for i:=1 to m do inp1?trial(b[i]);
      test(b, prime);
      inp2?complete;
      seed:=p+2
   until prime
end;
```

Algorithm 5.14: Generating a simple prime

```
procedure GenerateDoublePrime(t:integer; var r:integer;
   inp1,out1,inp2,out2:channel);
type table=array [1..m] of boolean;
var i,k:integer;
   b:table;
   prime:boolean;
begin
   k:=0;
   repeat
      out2!double(t, k);
      inp2?result(r);
      out2!complete;
      out1!certify(r);
      for i:=1 to m do inp1?trial(b[i]);
      test(b, prime);
      inp2?complete;
      k:=r div t
   until prime
end;
```

Algorithm 5.15: Generating a double prime

Algorithm 5.16 is the master procedure to generate a strong prime. It is identical to the two previous algorithms, except that a search is initiated for strong prime $p = p_0 + krs$, for some $k \geq 0$. If received candidate p is not prime, the process is repeated with a new search starting at $k = (p \text{ div } rs) + 2$. Since each of these three algorithms continues to search until a prime is found, this version of the master process will always generate a strong prime without failure.

```
procedure GenerateStrongPrime(r,s:integer; var p:integer;
    inp1,out1,inp2,out2:channel);
type table=array [1..m] of boolean;
var i,k:integer;
    rs,u,p0:integer;
    b:table;
    prime:boolean;
begin
    rs:=r*s
    u:=(rs+modpower(s,r-1,rs)-modpower(r,s-1,rs)) mod rs;
    if odd(u) then p0:=u else p0:=u+rs;
    k:=0;
    repeat
        out2!strong(p0, rs, k);
        inp2?result(p);
        out2!complete;
        out1!certify(p);
        for i:=1 to m do inp1?trial(b[i]);
        test(b, prime);
        inp2?complete;
        k:=p div rs
    until prime
end;
```

Algorithm 5.16: Generating a strong prime

```
procedure certifier(id:integer; inp,out:channel);
var x:integer;
    b,more:boolean;
    j,k,q:integer;
begin
    more:=true;
    while more do
        poll
            inp?certify(x) ->
                if id<p then out!certify(x);
                q:=m div p;
                for j:=1 to q do begin
                    solve(x, b);
                    out!trial(b);
                    for k:=1 to id-1 do begin
                        inp?trial(b);
                        out!trial(b)
                    end
                end |
            inp?eos ->
                if id<p then out!eos;
                more:=false
        end
end;
```

Algorithm 5.17: Modified primality testing server algorithm

```
procedure FindDouble(id:integer; x, k:integer;
   inp1,out1,out2:channel);
var y:integer;
   more,found:boolean;
begin
   k:=k+id*2;
   more:=true;
   while more do begin
      out1!check;
      poll
         inp1?complete -> more:=false |
         inp1?continue ->
            y:=k*x+1;
            found:=false;
            if not SmallFactor(y) then found:=prime(y, 1);
            if found then begin
               out2!result(y);
               out1!complete;
               more:=false
            end;
            k:=k+p*2
      end
   end;
   out2!complete
end;
```

Algorithm 5.18: *Modified search procedure to generate a double prime candidate*

The necessary modifications for the two sets of servers are straightforward. The primality testing servers must be modified to repeatedly accept input from the master: testing when a candidate is received, and terminating when an *eos* signal is received. Therefore, algorithm 5.17 replaces algorithm 5.9.

Algorithm 5.17 is a modification of the certifier procedure that repeatedly polls for either a *certify* symbol or an *eos* signal. If a *certify* symbol is received, the certification procedure of algorithm 5.9 is executed. When an *eos* signal is received, the certifier terminates.

Algorithm 5.18 is the search procedure for double primes. It is identical to the search procedure for simple primes, except that it tests numbers of the form $y = kx + 1$. To insure that each server tests a distinct set of numbers, each initializes local variable k to the value of $2id$ and increments k by $2p$.

Algorithm 5.19 is the search procedure for strong primes. It is identical to the search procedures for simple and double primes, except that it tests numbers of the form $y = p_0 + krs$. To insure that each server tests a distinct set of numbers, each initializes local variable k to the value of $2id$ and increments k by $2p$.

```
procedure FindStrong(id:integer; p0, rs, k:integer;
   inp1,out1,out2:channel);
var y:integer;
   more,found:boolean;
begin
   k:=k+id*2;
   more:=true;
   while more do begin
      out1!check;
      poll
         inp1?complete -> more:=false |
         inp1?continue ->
            y:=p0+k*rs;
            found:=false;
            if not SmallFactor(y) then found:=prime(y, 1);
            if found then begin
               out2!result(y);
               out1!complete;
               more:=false
            end;
            k:=k+p*2
      end
   end;
   out2!complete
end;
```

Algorithm 5.19: Modified search procedure to generate a strong prime candidate

5.12 Implementation Concerns

To construct a strong prime p of approximately n bits, we should generate the component primes r and s with approximately $n/2$ bits each. In general, it is difficult to generate a strong prime with an exact length of n bits. For applications where this is necessary, [Gordon 84] makes a few suggestions toward achieving this goal.

The integer arithmetic in the algorithms above must be replaced with *multiple-length arithmetic*, since the integers involved are far larger than the word-size on most computers. The programming of the multiple-length operations is straightforward, with the notable exception of multiple-length division [Brinch Hansen 92d].

We perform modular exponentiation, which is the heart of the Miller-Rabin witness test, using the well-known fast exponentiation technique of successive squaring [Bressoud 89]. We choose this approach because it is well-suited for testing quadratic residues, as is necessary for the Miller-Rabin algorithm. We should note, however, that there are many proposed algorithms for even faster modular exponentiation [Agnew 88, Beth 86, Bos 89, Findlay 89, Geiselmann 90, Ghafoor 89, Kawamura 88, Morita 89, Yacobi 90], many of which are specifically intended for a VLSI implementation.

5.13 Performance Analysis

Execution of the prime generation algorithm includes three time-consuming components: the time required for certification, the time required to generate prime candidates, and the time required to terminate each search for candidates (after a candidate has been found). We assume that the communication time for the algorithm is negligible.

Ideally, our parallel algorithm should achieve a speedup of p for the generation of prime candidates and the certification of primes, when executing on p processors. The termination time, however, is overhead. We, therefore, expect to have a parallel execution time of the form

$$T(p) = \frac{T_{certification} + T_{generation}}{p} + T_{termination}$$

From [Brinch Hansen 92b] we know that the execution time for the Miller-Rabin algorithm applied to an N-bit number is of the form $aN^3 + bN^2$, where a and b are system-dependent constants and a is significantly smaller than b.

The execution time for prime generation is the time required for trial division by small primes, together with the time required for Miller-Rabin witness tests. On average, we expect to test $O(\ln n)$ numbers close to n, before finding a prime. If n is an N-bit number, $O(\ln n) = O(N)$. Therefore, we expect to apply the trial division algorithm to $O(N)$ numbers before finding a candidate. A small fraction of those numbers will also require a Miller-Rabin test.

Trial division by a small prime (all of which require just a single digit in base-2^{32}) is an $O(N)$ operation. There is a constant limit on the number of trial divisions that may be performed for any number being tested, so the time spent on trial division for each candidate is just $O(N)$. Since $O(N)$ candidates are tested, we expect the total execution time for trial division to be $c_1 N^2$, where c_1 is a system-dependent constant.

The execution time for the Miller-Rabin witness tests during candidate generation is dependent upon the number of applications of the test. Again, we expect $O(N)$ numbers to pass the trial division test, and require a Miller-Rabin test. Since the execution time for each Miller-Rabin test is of the form $aN^3 + bN^2$, the total execution time for the witness tests executed during candidate generation should be $aN^4 + b_1 N^3$, where a and b_1 are system-dependent constants.

The time required to terminate a search for prime candidates depends on the test being executed by the servers that have not yet found a candidate. (Recall that a search process cannot recognize the search completion signal until it completes its test of the current number.) If all of the servers are testing numbers with small factors, then we would expect the termination time to be $O(N)$, which is negligible. However, if at least one of the servers is testing a number without small factors, that server will apply a Miller-Rabin witness test, and we would expect the termination time to be $O(N^3)$.

Empirical data shows that each candidate generating server spends the vast majority of its execution time performing witness tests. (This is a result of the fact that the witness test is much slower than trial division.) Therefore, when we are running in parallel, there will almost always be at least one server that must complete a witness test before terminating its search. Therefore, we expect the termination time to be $dN^3 + eN^2$, where e and d are system-dependent constants.

Finally, certification requires a constant number of Miller-Rabin tests. Therefore, we expect the certification time to be $b_2 N^3 + c_2 N^2$, where b_2 and c_2 are system-dependent constants.

Combining the system-dependent constants so that $b = b_1 + b_2$ and $c = c_1 + c_2$, the expected parallel execution time for generating N-bit strong primes on p processors (excluding the master) is

$$T(p,N) \approx \frac{aN^4 + bN^3 + cN^2}{p} + \left(dN^3 + eN^2\right), \quad \text{for } p > 1$$

where the constants are associated with the following operations:

a generation,
b generation and certification,
c trial division and certification,
d termination,
e termination.

For execution on a single server process, there is no termination overhead, so we have

$$T(1,N) \approx aN^4 + bN^3 + cN^2.$$

For our implementation in occam 2 on a Meiko Computing Surface with T800 transputers, we found the following values for the constants:

$$a = 0.0027 \ \mu s, \ b = 2.4 \ \mu s, \ c = 840 \ \mu s, \ d = 0.058 \ \mu s, \ e = 41 \ \mu s.$$

Currently, a minimum-security RSA cryptosystem requires a modulus of at least 512 bits. Such a modulus would require strong primes of approximately 256 bits each. A high-security RSA cryptosystem currently requires a modulus of approximately 660 bits, having strong prime factors of approximately 330 bits each [Brassard 88, Denning 82]. For applications where security is critical (for example, in military use), a modulus as large as 1024 bits might be used [Beth 91]. To test our parallel algorithm, we generated strong primes with slightly more than 256, 320, and 512 bits.

The running time for a sequential prime number generator is highly dependent upon the starting seed, since primes are not distributed uniformly. By chance, we may select a seed that is either very close to a prime, or very far from one. In addition, the non-deterministic nature of our parallel algorithm will cause the particular

Processors	T (s) 256 bits	T (s) 320 bits	T (s) 512 bits
1	107 (107)	186 (193)	725 (727)
5	24.7 (25.0)	42.7 (44.7)	158 (164)
10	13.9 (14.4)	25.1 (25.4)	92.7 (91.3)
20	8.77 (9.01)	15.4 (15.7)	57.8 (54.9)
40	6.22 (6.33)	10.6 (10.9)	37.7 (36.7)

Table 5.1: *Observed and predicted running times for strong prime generation*

prime generated (and consequently, the running time of the algorithm) to vary with the number of processors used.

Since approximately one in every $\ln n$ numbers around n is a prime, testing should be averaged over $O(\ln n)$ trials, ideally. We averaged our results for fifty different pairs of starting seeds. Table 5.1 shows the observed (and predicted) average running times for the fifty trials.

Figures 5.3 and 5.4 show the estimated·average speedup and parallel efficiency. (We plot the estimated figures based on the analytic model, since this reduces variations due to the limited number of trials.)

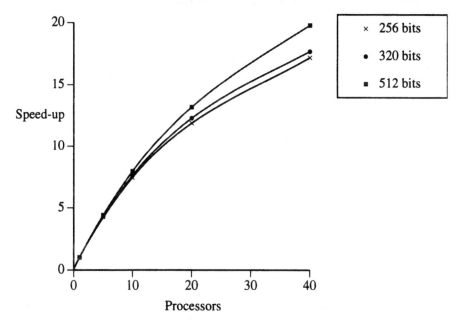

Figure 5.3: *Speedup for parallel strong prime generation*

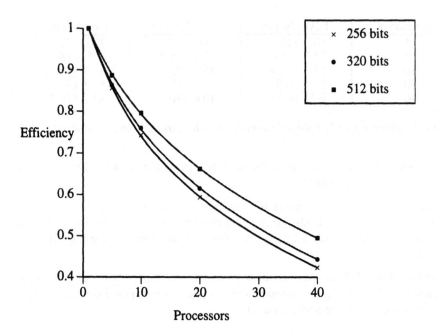

Figure 5.4: *Parallel efficiency of strong prime generation*

Chapter 6

Distributed Factoring with Competing Servers

"It is the mark of an inexperienced man not to believe in luck."

Joseph Conrad

6.1 Introduction

The RSA cryptosystem is based on modular exponentiation, and its security requires that it be infeasible to factor the RSA modulus. In order to make factoring an RSA modulus infeasible, the modulus is chosen to be the product of two large primes. This renders general-purpose factorization techniques useless. However, some products of two large primes are susceptible to factorization by one of a number of special-purpose factorization techniques.

This chapter describes one special-purpose factoring algorithm, the Pollard $p-1$ method and develops a distributed program to implement the method, based on the competing servers array. The distributed program can be easily modified to implement other special-purpose factoring algorithms, including the Pollard Rho, Williams $p+1$, and Elliptic Curve methods of factoring. As we did with the Pollard $p-1$ method, this is achieved by simple substitution of data types and a few sequential procedures.

In addition, the program that we have present can easily be modified to apply several of the methods simultaneously, simply by having different servers apply different factoring methods.

6.2 Special-Purpose Factoring Algorithms

Since factoring is intimately tied to the security of the RSA cryptosystem, factoring algorithms are of great concern when using the cryptosystem. Successfully factoring an enciphering modulus breaks that instance of the cryptosystem. Therefore, a cryptographer must insure that a modulus used for RSA cannot be factored (in a practical amount of time) using any algorithm.

While the largest number reported to have been factored using a general-purpose factoring algorithm had only 116 decimal digits [Silverman 91], much larger numbers (and even many numbers which are the product of two large primes) can be successfully factored using one of a number of special-purpose factoring techniques [Bressoud 89, Riesel 85]. These special-purpose factoring techniques are probabilistic algorithms that exploit special properties of the numbers they factor. As a result, they are effective only for products that have those special properties, and in general, they cannot be guaranteed to successfully factor any given product.

There are two main uses for special-purpose factoring algorithms. First, when one attempts to factor a large product, special-purpose factoring algorithms can be used as a first approach to factoring. If the relatively fast special-purpose techniques fail to produce any factors, one can revert to one of the much slower, general-purpose algorithms.

Another important use of special-purpose factoring algorithms is for testing RSA moduli to insure that they cannot be easily factored using one of the special-purpose factoring techniques. Applying the special-purpose algorithms, one can discover (and discard) moduli that are insecure under a special-purpose factoring attack.

There are many different special-purpose factoring algorithms. Two of the most important are *Pollard's $p-1$ method* [Pollard 74], and *Williams' $p+1$ method* [Williams 82]. The Pollard $p-1$ method can factor a modulus very quickly if the modulus has a factor, p, where all the factors of $p-1$ are small (say, less than ten thousand, or one million). Similarly, the Williams $p+1$ method can quickly factor a modulus with a factor, p, where all the factors of $p+1$ are small.

Two other common algorithms are the *Pollard Rho method* [Pollard 75] and the *Elliptic Curve method* [Bressoud 89, Lenstra 87]. The Elliptic Curve method is a generalization of the Pollard and Williams algorithms, and is considerably more general than the other algorithms in its applicability.

These factoring techniques are all *parameterized factoring algorithms* that are probabilistic in approach. They each use some parameter that may be arbitrarily varied in order to change the performance of the algorithm. It has been shown, both theoretically and empirically, that a special-purpose factoring attack is more likely to be successful when a number of short runs, each with a different choice of parameter, are used, than when the same amount of computing time is spent on one long run with a single choice of parameter [Bressoud 89].

As a result, these algorithms can be easily parallelized, simply by executing independent runs in parallel. Since we need only find a single factor (as we are interested in RSA moduli that are the product of just two primes), the competing servers program shell developed in chapter 4 is well-suited to implement special-purpose factoring as a distributed program.

As a sample special-purpose factoring technique, we shall examine Pollard's $p-1$ method. We choose Pollard's $p-1$ method because it is easy to understand.

The other special-purpose factoring techniques can be similarly implemented in parallel, by simple substitution of their sequential algorithms.

6.3 Pollard's $p-1$ Method

Pollard's $p-1$ method attempts to factor a product by constructing a second product that shares a non-trivial factor with the original product. (A trivial factor would be either 1 or the original product.) When this can be done, computing the greatest common divisor of the two products reveals the shared factor.

Pollard's method attempts to construct the new product by exploiting a property of primes defined by *Fermat's Theorem* [Niven 80].

Fermat's Theorem: If p is a prime that does not divide c, then $c^{p-1} \equiv 1 \pmod{p}$.

Suppose that we wish to factor integer n. Furthermore, suppose that n has an unknown prime factor p, such that all of the prime factors of $p-1$ are small. We consider the factors of $p-1$ to be small if $p-1$ divides $k!$, for some small value of k, say 10000.

Although p is unknown, we attempt to find another integer m, such that $m-1$ is divisible by p. Once we have n and $m-1$ divisible by p, we can compute p (or a multiple of p) by computing $\gcd(m-1, n)$.

Toward this end, we define

$$m = c^{10000!} \bmod n \qquad (6.1)$$

where c is a positive integer greater than 1 and less than p. We can compute m with reasonable speed since

$$c^{10000!} = \left(\cdots \left(\left(c^1 \right)^2 \right)^3 \cdots \right)^{10000}$$

and since modular exponentiation is reasonably fast (requiring only polynomial time).

Now we show that, if $p-1$ divides 10000!, $m-1$ will contain a factor of p. Since $p-1$ divides 10000!, there is some integer r such that $(p-1)r = 10000!$. Therefore, we may rewrite equation (6.1) as

$$m = c^{(p-1)r} \bmod n$$

$$= \left(c^{p-1} \bmod n \right)^r \bmod n$$

Taking m modulo p we get,

$$m \bmod p = \left(\left(c^{p-1} \bmod n\right)^r \bmod n\right) \bmod p$$

$$= \left(c^{p-1} \bmod p\right)^r \bmod p \qquad \textit{since n is a multiple of p}$$

$$= 1^r \bmod p \qquad \textit{by Fermat's theorem,}$$
since $c < p$, p is a prime that
does not divide c

$$= 1$$

Which simply means that $m - 1$ is a multiple of p.

Now we have two numbers with p as a factor: n and $m - 1$. If $m - 1$ does not contain all of the other factors of n (that is, if $m - 1$ is not a multiple of n), then

$$g = \gcd(m - 1, n)$$

is a non-trivial factor of n.

In essence, Pollard's method is to compute

$$g = \gcd\left(\left(c^{k!} \bmod n\right) - 1, n\right),$$

where k is a small positive integer, in the hope that g will be a non-trivial divisor of n. In practice, we do not know how large k must be before $(c^{k!} \bmod n) - 1$ will be a multiple of a prime factor of n. Furthermore, if k is too large, then $(c^{k!} \bmod n) - 1$ may be a multiple of all the prime factors of n. That is, it will be a multiple of n, and the gcd will be trivial divisor n.

Example: Consider the number

$$n = 42248651.$$

Applying the Pollard method with $c = 2$ and $k = 31$, we find that

$$\gcd\left(\left(2^{31!} \bmod n\right) - 1, n\right) = \gcd(28311279, 42248651)$$

$$= 7937$$

Indeed, it turns out that

$$n = 42248651$$

$$= 5323 \times 7937$$

where

$$5323 = 2 \times 3 \times 887 + 1$$

and

$$7937 = 2^8 \times 31 + 1$$

are primes.

Since 7936 divides 31! but 5322 does not, $\gcd\left((2^{31!} \bmod n) - 1, n\right)$ yields the prime factor 7937.

If we had chosen $k \geq 887$, both 7936 and 5322 would have divided $k!$ and $(2^{k!} \bmod n) - 1$ would have been a multiple of both 7936 and 5322. Therefore, $\gcd\left((2^{k!} \bmod n) - 1, n\right)$ would yield trivial factor n.

End Example.

To convert this method into a practical algorithm, we simply compute $\gcd\left((c^{k!} \bmod n) - 1, n\right)$ for increasing values of k, until we find a non-trivial divisor, or until we reach some limiting value for k (for example, 10,000).

Algorithm 6.1 is a procedure for Pollard's $p-1$ method. The procedure uses constant *limit* to denote the limiting value for k. Parameter n is the integer to be factored. c is the base to be used for factoring. The factor found is assigned to variable parameter g. If a non-trivial factor is not found, the value of g will be trivial factor n.

Since we compute $m = c^{k!} \bmod n$ for increasing values of k, we can think of m as a function of k, m_k. If we know the value of m_{k-1}, then we can compute m_k as

$$
\begin{aligned}
m_k &= c^{k!} \bmod n & &\text{\textit{by definition of } } m_k \\
&= c^{k(k-1)!} \bmod n & &\text{\textit{by definition of } } k! \\
&= \left(c^{(k-1)!} \bmod n\right)^k \bmod n \\
&= \left(m_{k-1}\right)^k \bmod n. & &\text{\textit{by definition of } } m_{k-1}
\end{aligned}
$$

The algorithm initializes variable m to hold the value of $m_1 = c$. The algorithm then proceeds to compute m_k for increasing values of k. For each value of k, we may compute the next value of m as the last value of m raised to the power k and taken modulo n.

```
procedure factor(n,c:integer; var g:integer);
var k,m:integer;
begin
   m:=c; k:=2; g:=1;
   while (k<=limit) and (g=1) do begin
      m:=modpower(m, k, n);
      g:=gcd(m-1, n);
      k:=k+1
   end
end;
```

Algorithm 6.1: *Pollard's p–1 factoring method*

The operation *modpower(m, k, n)* computes the value of m^k mod n. The operation *gcd(m–1, n)* computes the gcd of $m - 1$ and n.

6.4 Parallel Factoring

Our basic approach to factoring in parallel is to allow multiple server processes to execute independent runs of Pollard's $p - 1$ method, each using a distinct base c. Each server process will execute Pollard's $p - 1$ method (possibly executing several runs) until either a factor is found, or the factoring attempt is terminated. Our distributed factoring algorithm will use a variant of the competing servers generic program that allows a computation to be timed out by the master process [Chapter 4].

The generic program for competing servers uses a channel protocol defined by the type declaration

```
channel = [start(SearchDataTypes), result(ResultTypes),
    check, continue, complete, eos];
```

We must replace *SearchDataTypes* and *ResultTypes* with data types appropriate for our particular application.

The core of computation in the competing servers shell is the *find* procedure executed by a *searcher* subprocess of the server process, and defined in chapter 4. The procedure defines a general search that is periodically stopped to check whether the search should be terminated (as a result of some other server having found a solution, or as a result of the computation being timed out).

```
procedure FindFactor(id:integer; n:integer;
    inp1,out1,out2:channel);
var g,c,m,k:integer;
    more,found:boolean;
begin
    initialize(c, m, k, g, id);
    more:=true;
    while more do begin
        out1!check;
        poll
            inp1?complete -> more:=false |
            inp1?continue ->
                factor(n, m, k, g, found);
                if found then begin
                    out2!result(g);
                    out1!complete;
                    more:=false
                end;
                update(n, c, m, k, g)
        end
    end;
    out2!complete
end;
```

Algorithm 6.2: *Modified find procedure for Pollard p–1 factoring*

```
procedure initialize(var c,m,k,g:integer; id:integer);
begin
    c:=id+1;
    m:=c;
    k:=1;
    g:=1
end;
```

Algorithm 6.3: *Initialize procedure for Pollard's p–1 factoring method*

Algorithm 6.2 is a modified version of the *find* procedure, specialized to search for factors using Pollard's $p-1$ method. In order to factor, we must start with a number to be factored. Once a factor is found, it is returned as the result. Therefore, we substitute the type *integer* for both *SearchDataTypes* and *ResultTypes*.

Parameter n is the number to be factored. *id* is a process number. The servers are numbered 1 to p, where p is a constant. Each server must attempt to factor using distinct base constants. To achieve this, each server initializes its base constant to $id+1$. For any subsequent factoring attempts, the server increments the base constant by p. This interleaves the base constants used by each server.

The parameters *inp1* and *out1* are channels connecting the searcher process to a local *reader* process. During the search for factors, each searcher periodically communicates with its local reader. The searcher sends a *check* signal to the reader. The reader responds with either a *complete* signal or a *continue* signal. A *complete* signal indicates that the search should be terminated. A *continue* signal indicates that the search should be continued [Chapter 4].

out2 is a channel connecting the *searcher* process to a local *writer* process. If the searcher finds a factor, it outputs that factor to its local writer.

Algorithm 6.3 is the *initialize* procedure. The procedure implements the necessary initialization of the base constant c, the value m, the index k, and the gcd, g.

Algorithm 6.4 is the *factor* procedure. It implements one step of the factoring method and is derived from algorithm 1 in a straightforward manner.

The *update* procedure is algorithm 6.5. If the limiting value for k has been reached, or if the factoring method has found only trivial divisor n, the procedure increments base constant c and reinitializes the other variables to start a new factoring attempt.

```
procedure factor(n:integer; var m,k,g:integer;
    var found:boolean);
begin
    m:=modpower(m, k, n);
    g:=gcd(m-1, n);
    k:=k+1;
    found:=(g>1) and (g<n)
end;
```

Algorithm 6.4: *Factor procedure for Pollard's p–1 factoring method*

```
procedure update(n:integer; var c,m,k,g:integer);
begin
    if (k>limit) or (g<>1) then begin
        c:=c+p;
        m:=c;
        k:=1;
        g:=1
    end
end;
```

Algorithm 6.5: *Update procedure for Pollard's p−1 factoring method*

Since we want to be able to terminate the factoring computation if it fails to find a factor after some specified amount of time, we use the timeout variation of the master process in chapter 4. Algorithm 6.6 is the master procedure with the necessary substitutions for use in the factoring program.

The master initiates the search for factors by broadcasting the number to be factored. Subsequently, the master waits until either a factor is returned or a pre-defined execution time limit is exceeded. If a factor is returned, the master broadcasts a *complete* signal to terminate the search.

If no factor is returned within the execution time limit, the master outputs a *complete* signal to terminate the search. Subsequently, the array of servers may return either a factor or a *complete* signal, and the master must poll its input channel accordingly [Chapter 4].

Finally, the master broadcasts an *eos* signal to terminate the servers.

Together with the remainder of the competing servers program shell described in chapter 4, this defines a complete distributed algorithm for Pollard's $p-1$ factoring method.

```
procedure master(n,timeLimit:integer; var g:integer;
    var found:boolean; inp,out:channel);
begin
    out!start(n);
    found:=true;
    poll
        inp?result(g) ->
            out!complete;
            inp?complete |
        after timeLimit ->
            out!complete;
            poll
                inp?result(g) -> inp?complete |
                inp?complete -> found:=false
            end
    end;
    out!eos
end;
```

Algorithm 6.6: *Master procedure for factoring*

6.5 Implementation Concerns

For the sake of simplicity, we have described a basic implementation of Pollard's $p-1$ method. There are several optimizations to the basic algorithm that can be used to improve the performance of the algorithm. These optimizations can reduce the execution time of the algorithm by an order of magnitude [Bressoud 89].

The integer arithmetic in the algorithms above must be replaced with *multiple-length arithmetic*, since the integers involved are far larger than the word-size on most computers. The programming of the multiple-length operations is straightforward, with the notable exception of multiple-length division [Brinch Hansen 92d].

6.6 Performance

We reprogrammed the distributed factoring program using occam 2 on a Meiko Computing Surface with T800 transputers. To test the correctness of our program, we used it to factor 640 bit products primes p and q, where $p-1$ and $q-1$ were constructed from small factors. For $p-1$ and $q-1$ constructed from factors less than 500 (using the algorithm in chapter 9), the program required from three to ten minutes to extract 320 bit prime factors.

We then used the program to verify the security of RSA moduli against a $p-1$ factoring attack. For moduli of 640 bits, and an index limit of 10,000 for k, we found that a single processor could perform one run of the Pollard $p-1$ method in approximately ninety minutes. Not surprisingly, with forty server processes, the distributed factoring program could perform forty runs of the method, in the same amount of time.

Chapter 7

The Synchronized Servers Pipeline

"The best way out is always through."

Robert Frost

7.1 Introduction

This chapter describes a distributed program shell which implements a programming paradigm called the *synchronized servers*. The synchronized servers paradigm is related to the Monte Carlo paradigm [Brinch Hansen 92c]. The Monte Carlo paradigm includes problems where a probabilistic computation is repeatedly performed with the same input. The synchronized servers paradigm, on the other hand, includes problems where a particular computation is repeated with different input data.

In both cases, the computations may be performed by multiple server processes working independently and in parallel. However, while the Monte Carlo paradigm requires a simple broadcast of a single input to all server processes, the synchronized servers paradigm requires the distribution of a different inputs to each server process. Therefore, the paradigm includes problems suitable for a simple *domain decomposition* or *data parallel* approach [Cok 91, Fox 88].

A generic program for the paradigm uses multiple server processes that work in lock-step synchronization. When using a pipeline architecture, the distribution of input data and the collection of output results can be implemented in shift-register fashion, with data and results shifted through the pipeline. Since the servers all perform input, computation, and output at the same time, their behavior is highly synchronized.

The program shell for the synchronized servers paradigm can be used to implement any application which requires that a number of identical, independent computations be performed on multiple inputs. (Of course, to achieve efficient performance, the computation time must be large enough to make the communication time negligible.) We shall use our program shell to implement *parallel enciphering* and *deciphering* for the *RSA cryptosystem* [Rivest 78b, Chapter 2]. We shall additionally use the program shell to implement a *deterministic primality testing* algorithm.

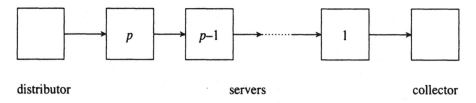

distributor   servers   collector

Figure 7.1: *Synchronized servers pipeline with distributor and collector*

7.2 The Synchronized Servers Pipeline

A distributed implementation of the synchronized servers paradigm uses a number of server processes executing in parallel. A distributor process distributes data to the server processes, and a collector process collects results from the processes.

Each server process must be able to receive input from the distributor process. It must also be able to send output to the collector process. A server process has no need to interact directly with any other server process, however. Naturally, the distributor process must be able to send output to each of the server processes, and the collector process must be able to receive input from each of the servers.

Of the available choices for implementing the synchronized servers, we choose to connect them as a *pipeline*. We choose a pipeline because the communication time will be negligible compared to the computing time (at least for the problems of RSA enciphering and primality testing), and because the pipeline implementation leads to a very regular, shift-register-like communication pattern.

Figure 7.1 shows a pipeline of servers numbered right-to-left from 1 to p, where p is a constant. Since the distributor and collector processes are not directly connected to every server process, the server processes have to perform message routing.

Figure 7.2 identifies the connections for a single server process in the array. For any server process in the pipeline, the adjacent processes on the left and right are called the *predecessor* and *successor*, respectively. Depending upon the position of the server process within the array, the predecessor may be either another server process, or the distributor process. Similarly, the successor may be either another server process or the collector process.

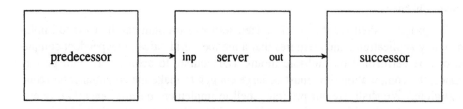

Figure 7.2: *Detailed view of server process connections*

7.3 Implementing Synchronized Servers

A server process attempts to execute four phases in the course of a computation. First, there is a *broadcast phase*, in which global information needed by all servers is broadcast. Next there is a *shift-in* phase, in which input data is shifted into the server pipeline. Subsequently, there is a *computation phase*, in which the servers compute results in parallel. Finally, there is *shift-out phase*, in which the computational results are shifted out of the server pipeline.

In general, the number of computations may not equal the number of servers in the pipeline. If the number of servers exceeds the number of computations, some of the servers will have no input data upon which to perform the sequential computation, and will be *idle*. An idle server does not execute the computation and shift-out phases.

On the other hand, if the number of computations exceeds the number of servers, the computations cannot all be performed during the execution of the four phases. In this case, all execution phases, except broadcasting, are repeated until the required number of computations have been performed.

We call the shift-in, computation and shift-out phases a *step* in the execution of the synchronized servers. A step in which no server is idle is called a *complete* step. A step in which at least one server is idle is called an *incomplete* step.

In general, performing an arbitrary number of computations will require zero or more complete steps, followed by a final incomplete step. For example, performing forty-seven computations on a pipeline of ten synchronized servers would require four complete steps, followed by one incomplete step. During the incomplete step, three of the servers would be idle.

The final incomplete step may not require any computations. For example, performing ten computations on a pipeline of ten synchronized servers requires one complete step followed by one incomplete step. During the incomplete step, no computations are performed. The step merely involves the transfer of the completion signal through the pipeline.

We start the definition of our program shell by specifying a channel protocol for the communicating processes. The type definition

```
channel = [global(GlobalDataTypes), data(InputTypes),
    result(ResultTypes), complete, eos];
```

is used throughout the program, and is presumed to have been previously declared. We shall describe the meaning of each message as we use them.

Our algorithm is designed to support applications where the number of computations required may not be known in advance. Therefore, input is sent to the server pipeline as a sequence of *data* symbols followed by a completion signal, *complete*.

Algorithm 7.1 defines a server process. Parameter *id* is a process number that identifies a server's position within the pipeline. Channel parameters *inp* and *out* link

```
procedure server(id:integer; inp,out:channel);
var g:GlobalDataTypes;
    a:InputTypes;
    b:ResultTypes;
    more:boolean;
begin
    broadcast(g, id, inp, out);
    more:=true;
    while more do begin
        ShiftIn(a, more, id, inp, out);
        if more then begin
            compute(g, a, b);
            ShiftOut(b, more, id, inp, out)
        end
    end
end;
```

Algorithm 7.1: The server procedure

a server to its predecessor and successor, respectively.

A server process executes a separate procedure during each phase of execution. Initially there is a broadcast phase, during which global data for the computation is assigned to variable g.

Subsequently, a server process attempts to execute a step. If the step is incomplete and the server is idle, the boolean variable *more* is assigned the value of false. The idle server then terminates.

If the server is not idle, the data for the local computation is assigned to variable a. The server proceeds to perform the computation phase, producing result b. Finally, the server outputs the result during the shift-out phase. If a completion signal is received during the shift-out phase (which happens when the step is incomplete), *more* is assigned the value of false, and the server terminates. Otherwise, the server attempts to execute another step.

Algorithm 7.2 is the broadcast procedure. During the broadcast phase, the distributor outputs global data to the servers using a *global* message. Each server inputs the data from its predecessor and forwards the data to its successor, unless it is the last server on the pipeline. The global data is assigned to variable parameter g.

The shift-in procedure is algorithm 7.3. The input channel is repeatedly polled for either input data or the completion signal.

```
procedure broadcast(var g:GlobalDataTypes; id:integer;
    inp,out:channel);
begin
    inp?global(g);
    if id>1 then out!global(g)
end;
```

Algorithm 7.2: The broadcast procedure

```
procedure ShiftIn(var a:InputTypes; var more:boolean;
    id:integer; inp,out:channel);
var i:integer;
begin
    i:=1;
    more:=true;
    while more and (i<id) do
        poll
            inp?data(a) ->
                out!data(a);
                i:=i+1 |
            inp?complete ->
                out!complete;
                more:=false
        end;
    if more then
        poll
            inp?data(a) -> |
            inp?complete ->
                out!complete;
                more:=false
        end
end;
```

Algorithm 7.3: The shift-in procedure

Let us first consider execution of the procedure during a complete step. During the shift-in phase of a complete step, each server process must transfer enough data to supply all of the servers that succeed it. $id - 1$ servers succeed a server with process number id (see Fig. 7.1). As a result, each server's first action during the shift-in phase is to copy $id - 1$ *data* symbols from its input channel to its output channel. Once this has been done, the server inputs one more *data* symbol from its predecessor. The data contained in that message is assigned to variable parameter a.

Now, let us consider execution during an incomplete step. In total, server number id attempts to input id *data* symbols during the shift-in phase. If a completion signal is received before id *data* symbols have been received, the server will forward the signal to its successor, terminate the shift-in phase, and update variable parameter *more* to indicate that execution should be terminated.

The generic program is adapted to solve a particular problem by defining a *compute* procedure with an appropriate application-specific sequential computation.

Algorithm 7.4 is the shift-out procedure. A server's first action during the shift-out phase it to output a *result* symbol containing local result b. Afterward, it must forward any results from the servers that precede it on the pipeline. For the forwarding process, the server polls the input channel for either results or a completion signal.

Server number id is preceded on the pipeline by $p - id$ server processes (see Fig. 7.1). Therefore, after outputting its own result, a server attempts to copy $p - id$ *result* symbols from its input channel to its output channel. If a completion signal is

```
procedure ShiftOut(b:ResultTypes; var more:boolean;
    id:integer; inp,out:channel);
var i:integer;
begin
    out!result(b);
    i:=id+1;
    more:=true;
    while more and (i<=p) do
        poll
            inp?result(b) ->
                out!result(b);
                i:=i+1 |
            inp?complete ->
                out!complete;
                more:=false
        end
end;
```

Algorithm 7.4: The shift-out procedure

received, the phase is terminated, and variable parameter *more* is updated to indicate that the computation has been completed.

7.4 Distribution and Collection

The particular behavior of the distributor and collector processes is dependent upon the specific use of the server pipeline. The distributor and collector simply provide an interface for the synchronized servers.

In some cases, the distributor and collector may interface with a filing system. In that case, the distributor might read data from a file, and output it to the server pipeline, as shown in algorithm 7.5. Similarly, the collector might write the results to a file, as shown by algorithm 7.6.

In other cases, the distributor and collector may act as an interface to other processes. For example, our primality testing implementation based on synchronized servers will be one part of a larger distributed program. In that situation, the distributor and collector are just buffer processes.

```
procedure distributor(out:channel);
var g:GlobalDataTypes;
    a:InputTypes;
begin
    ReadGlobalData(g);
    out!global(g);
    while not eof do begin
        ReadData(a);
        out!data(a);
    end;
    out!complete
end;
```

Algorithm 7.5: A distribution procedure

```
procedure collector(inp:channel);
var b:ResultTypes;
    more:boolean;
begin
    more:=true;
    while more do
        poll
            inp?result(b) -> WriteData(b) |
            inp?complete -> more:=false
        end
end;
```

Algorithm 7.6: *A collection procedure*

```
procedure SynchronizedServers;
type net=array [0..p] of channel;
var c:net;
    i:integer;
begin
    parbegin
        distributor(c[p]) |
        collector(c[0]) |
        parfor i:=1 to p do
            server(i, c[i], c[i-1])
    end
end;
```

Algorithm 7.7: *The synchronized servers array*

The distributor and collector processes execute in parallel with the server pipeline, as defined by algorithm 7.7.

7.5 I/O Traces

The notation $\langle a_r...a_s \rangle$ denotes a sequence of symbol values a_r through a_s, transferred over a channel. (We ignore symbol names.) If $r > s$, then by definition $\langle a_r...a_s \rangle$ denotes an empty sequence. Using this notation, we can make assertions about the inputs and outputs of data values to examine the overall behavior of the pipeline.

The input to the server pipeline is the sequence $\langle g \rangle \langle a_1...a_n \rangle$ consisting of global data g followed by a sequence of input data. The corresponding output from the pipeline is the sequence of results $\langle b_1...b_n \rangle$, where $b_i = f(g, a_i)$, for some application-specific function f.

To see that this is true, let us first consider a pipeline of p servers and an ordered input sequence of p data values, $\langle a_1...a_p \rangle$. (We ignore the broadcast of global data in the remainder of this discussion.) Processing these inputs requires a complete step.

During a complete step, only data symbols are input by each server. Therefore, the body of the *ShiftIn* procedure of algorithm 7.3 is equivalent to the program segment:

```
for i:=1 to id-1 do begin
    inp?data(a);
    out!data(a);
end;
inp?data(a)
```

During the shift-in phase of a complete step, server number *id* inputs the sequence $\langle a_1 \ldots a_{id} \rangle$ and outputs the sequence $\langle a_1 \ldots a_{id-1} \rangle$.

During a complete step, the body of the *ShiftOut* procedure of algorithm 7.4 is equivalent to the program segment:

```
out!result(b)
for i:=id+1 to p do begin
    inp?result(b);
    out!result(b);
end;
```

During the shift-out phase, server number *id* inputs $\langle b_{id+1} \ldots b_p \rangle$ and outputs $\langle b_{id} \ldots b_p \rangle$.

The servers are numbered from 1 to *p*. The leftmost server is number *p* and the rightmost server is number 1 (see Fig. 7.1). During the shift-in phase, the leftmost server inputs

$$\langle a_1 \ldots a_{id} \rangle = \langle a_1 \ldots a_p \rangle$$

from the distributor process. The rightmost server outputs

$$\langle a_1 \ldots a_{id-1} \rangle = \langle a_1 \ldots a_0 \rangle,$$

which is the empty sequence, to the collector.

During the shift-out phase, the leftmost server inputs

$$\langle b_{id+1} \ldots b_p \rangle = \langle b_{p+1} \ldots b_p \rangle,$$

which is the empty sequence, from the distributor. The rightmost process outputs

$$\langle b_{id} \ldots b_p \rangle = \langle b_1 \ldots b_p \rangle$$

to the collector.

Overall then, output of $\langle a_1 \ldots a_p \rangle$ from the distributor results in the input of $\langle b_1 \ldots b_p \rangle$ by the collector.

Now, let us consider a pipeline of *p* servers and an ordered input sequence of *n* data values, $\langle a_1 \ldots a_n \rangle$, where $n = qp + r$, *q* and *r* are positive integers, and $r < p$. In this case q complete steps followed by one incomplete step are necessary.

Since the servers can repeatedly execute complete steps as described above, output of

$$\langle a_1 \ldots a_{qp} \rangle = \langle a_1 \ldots a_p \rangle \langle a_{p+1} \ldots a_{2p} \rangle \cdots \langle a_{(q-1)p+1} \ldots a_{qp} \rangle$$

from the distributor results in the input of

$$\langle b_1 \ldots b_{qp} \rangle = \langle b_1 \ldots b_p \rangle \langle b_{p+1} \ldots b_{2p} \rangle \cdots \langle b_{(q-1)p+1} \ldots b_{qp} \rangle$$

by the collector.

We, therefore, examine the incomplete step required to process the last r data values, $\langle a_{qp+1} \ldots a_{qp+r} \rangle$. As only the order (and not the numbering) of the values is important, we can renumber the input values to be $\langle a_1 \ldots a_r \rangle$ to simplify our discussion.

Since $r < p$, there are fewer data values than server processes. As a result, servers 1 through r will perform computations, while servers $r+1$ through p will be idle.

For each idle server, the body of the *ShiftIn* procedure is equivalent to the program segment:

```
for i:=1 to r do begin
    inp?data(a);
    out!data(a);
end;
inp?complete;
out!complete;
```

During this shift-in phase, each idle server inputs the sequence $\langle a_1 \ldots a_r \rangle$ and outputs the same sequence. Subsequently, each idle server terminates its shift-in phase and performs neither the computation nor shift-out phases.

Servers 1 through r behave as a pipeline of r servers processing r data values. The input of $\langle a_1 \ldots a_r \rangle$ results in the corresponding output of $\langle b_1 \ldots b_r \rangle$.

Overall, output of the data sequence

$$\langle a_1 \ldots a_n \rangle = \langle a_1 \ldots a_p \rangle \langle a_{p+1} \ldots a_{2p} \rangle \cdots \langle a_{(q-1)p+1} \ldots a_{qp} \rangle \langle a_{qp+1} \ldots a_{qp+r} \rangle$$

by the distributor process results in the input of

$$\langle b_1 \ldots b_n \rangle = \langle b_1 \ldots b_p \rangle \langle b_{p+1} \ldots b_{2p} \rangle \cdots \langle b_{(q-1)p+1} \ldots b_{qp} \rangle \langle b_{qp+1} \ldots b_{qp+r} \rangle$$

by the collector process.

```
procedure server(id:integer; inp,out:channel);
var g:GlobalDataTypes;
    a:InputTypes;
    b:ResultTypes;
    continue,more:boolean;
begin
    broadcast(g, continue, id, inp, out);
    while continue do begin
        more:=true;
        while more do begin
            ShiftIn(a, more, id, inp, out);
            if more then begin
                compute(g, a, b);
                ShiftOut(b, more, id, inp, out)
            end
        end;
        broadcast(g, continue, id, inp, out);
    end
end;
```

Algorithm 7.8: The modified server procedure

7.6 Repeated Use of the Pipeline

In some cases, it may be desirable to use the synchronized servers repeatedly. For example, a primality testing pipeline may be used to test many numbers for primality. We can easily modify our servers to support this.

To facilitate repeated use of the pipeline, we introduce the *eos* signal. This allows us to differentiate between the completion of one computation and the termination of the server processes. The *complete* symbol signals the completion of one computation, while the *eos* symbol signals termination of the pipeline.

Algorithm 7.8 defines the procedure for a modified server process. A simple loop is used to perform repeated computations. The server process determines whether another computation is to be performed during the broadcast phase. The shift-in, computation and shift-out phases remain unchanged.

The procedure executed during the broadcast phase is modified to that of algo-

```
procedure broadcast(var g:GlobalDataTypes; var more:boolean;
    id:integer; inp,out:channel);
begin
    poll
        inp?global(g) ->
            if id>1 then out!global(g);
            more:=true |
        inp?eos ->
            out!eos;
            more:=false
    end
end;
```

Algorithm 7.9: The broadcast procedure

Processors	Throughput (Kbit/s)	Speedup	Efficiency
1	1.47	1.00	1.00
5	7.34	4.98	1.00
10	14.6	9.93	0.99
20	29.2	19.8	0.99
40	58.0	39.3	0.98

Table 7.1: *Throughput, speedup and efficiency for parallel RSA enciphering*

rithm 7.9. The input channel is polled for either global data or an *eos* signal. If global data is received, a execution proceeds as described earlier. If a termination signal is received, the server forwards the signal to its successor, and then terminates.

7.7 Performance

We reprogrammed the generic program for synchronized servers using occam 2 on a Meiko Computing Surface with T800 transputers, and used it to implement distributed programs for RSA enciphering and deciphering [Chapter 8], and for deterministic primality testing [Chapter 9].

Table 7.1 shows sample throughput, speedup and efficiency figures for distributed RSA enciphering with a modulus of 640 bits and a small enciphering exponent of 32 bits (for fast enciphering). The program is capable of achieving nearly-perfect speedup, running with 98 percent efficiency on forty processors, for the sample shown. See chapter 8 for a more detailed analysis of the performance of the algorithm.

Chapter 9 describes the performance of a prime generation program that performs primality testing using the synchronized servers pipeline.

Chapter 8

Message Block Chaining for Distributed RSA Enciphering

"Half the work that is done in this world is to make things appear what they are not."

Elias Root Beadle

8.1 Introduction

Despite the advantages of a public-key cryptosystem over a secret-key cryptosystem, the RSA cryptosystem has been slow to gain acceptance for practical use. This is primarily due to the computational intensity of RSA enciphering, as compared to typical secret-key cryptosystems.

Since speed is the main barrier to the use of RSA in practical systems, numerous simplifications and optimizations have been proposed in order improve the speed of RSA enciphering [Cooper 90, Hoornaert 88, Jung 87].

For many cryptosystems, including RSA, enciphering is a process of breaking a large message into small blocks and enciphering the individual blocks. A particular method for dividing a large message into small blocks is called a *mode of operation*. One natural approach to improve the speed of RSA enciphering is to encipher many blocks in parallel.

Unfortunately, the mode of operation normally used with an RSA cryptosystem prevents such a parallel implementation of RSA enciphering. We propose a new mode of operation for RSA enciphering. The proposed mode of operation preserves many of the important properties of the conventional mode, but also facilitates a parallel implementation of RSA enciphering. We then develop a distributed algorithm for RSA enciphering using a generic program for the synchronized servers paradigm [Chapter 7].

8.2 Modes of Operation

When applying many encryption techniques to long messages, there is a need to break the plaintext message into short blocks for enciphering. For example, an RSA enciphered message is encoded modulo M. Such an RSA enciphering can encode no more than M distinct messages. An arbitrary plaintext message to be enciphered, however, may contain considerably more information than could be encoded without loss, by a single application of the RSA enciphering algorithm.

To prevent any information loss, a long plaintext message is broken into short blocks, each of a size that may be encoded using the desired enciphering technique, without loss. The plaintext message becomes a sequence of plaintext message blocks and the enciphered message becomes a sequence of ciphertext blocks.

A variety of modes of operation exist for breaking a large message into short blocks. The simplest mode of operation is known as *Electronic Code Book (ECB)*. ECB consists of splitting the plaintext message into a sequence of short blocks, $m_0, m_1, m_2, \ldots$, and subsequently enciphering each of the short blocks, independently, to form a sequence of ciphertext blocks, $c_0 = E(m_0), c_1 = E(m_1), c_2 = E(m_2), \ldots$.

While this mode of operation is particularly simple to understand and implement, it suffers from a number of security problems when used for certain applications. In particular, ECB enciphering has the undesirable property that duplicate portions of plaintext may be encoded into duplicate ciphertext portions, with a relatively high probability. This can make cryptanalysis of the ciphertext considerably easier. In addition, since the ECB enciphering of a block is independent of its position within the plaintext message, a cryptanalyst may be able to cut and paste segments of ciphertext, in order to forge a message [Brassard 88].

A second mode of operation is known as *Cipher Block Chaining*. This is the mode of operation normally used for RSA enciphering [Brassard 88]. Like ECB, cipher block chaining starts with splitting the plaintext message into a sequence of short blocks, $m_0, m_1, m_2, \ldots$. The first plaintext block is then encoded to form ciphertext block $c_0 = E(m_0)$. The remaining blocks are encoded using a chaining technique, so that the second plaintext block is encoded to form ciphertext block $c_1 = E(c_0 \oplus m_1)$, where $\oplus$ represents bitwise exclusive-or. The third plaintext block is enciphered to form ciphertext block $c_2 = E(c_1 \oplus m_2)$, and in general, plaintext block m_i is encoded to form ciphertext block $c_i = E(c_{i-1} \oplus m_i)$, for $i > 0$.

Cipher block chaining possesses a number of properties that make it a desirable mode of operation. Duplicate portions of plaintext, encoded using this technique, have a very low probability of being encoded into duplicate portions of ciphertext. In addition, due to the chaining of ciphertext blocks, each ciphertext block depends upon all previous blocks of the message. This makes it virtually impossible for a cryptanalyst to simply cut and paste blocks of ciphertext in order to form a forged message.

At the same time, arbitrary blocks of ciphertext can be deciphered individually, without the need to decipher the entire ciphertext message. For $i > 0$,

$m_i = c_{i-1} \oplus D(c_i)$, so only one deciphering operation is required.

Similarly, cipher block chaining allows the deciphering algorithm to recover from temporary transmission or enciphering errors. For example, if ciphertext block c_i is improperly transmitted, only message blocks $m_i = c_{i-1} \oplus D(c_i)$ and $m_{i+1} = c_i \oplus D(c_{i+1})$, depend on the value of c_i. Therefore, only those two blocks are affected by the transmission error.

As a result, cipher block chaining maintains much of the simplicity of ECB, while providing more secure encoding [Brassard 88]. Other modes of operation may be used, but will not be discussed here.

8.3 Strategy for Parallelization

Cooper [90] proposed a parallel algorithm based upon low-level parallelization of multiplication. We wish to consider an algorithm based upon parallelism introduced at a higher level. Specifically, we consider an algorithm where multiple blocks of plaintext are simultaneously enciphered into the corresponding blocks of ciphertext. (We shall discuss the algorithm in terms of the enciphering process; however, due to the symmetry of RSA enciphering and deciphering, the results will apply equally well to the RSA deciphering process.)

If we consider RSA enciphering in the simplest possible context, using the ECB mode of operation, a parallel algorithm along these lines is simple to derive. Using ECB, the splitting operation is trivial, and the process of enciphering any given block of plaintext is completely independent of the remainder of the plaintext. As a result, it is possible to efficiently encipher the plaintext blocks in parallel. This approach to parallel enciphering is suggested by Er [91] and Yun [90].

Unfortunately, ECB should be avoided. In practice, RSA enciphering should be performed using the more secure cipher block chaining mode [Brassard 88]. It doesn't take long to see, however, that cipher block chaining serializes the enciphering of blocks. Recall that with cipher block chaining, plaintext block m_i is enciphered into ciphertext block $c_i = E(c_{i-1} \oplus m_i)$, where c_{i-1} is the previous block of ciphertext. Clearly then, the process of enciphering block m_i can not commence until previous plaintext block m_{i-1} has been enciphered. So, using cipher block chaining renders our parallel algorithm useless. What we need is a new mode of operation that possesses the desirable properties of cipher block chaining, yet allows the enciphering of blocks to be performed in parallel, like ECB.

8.4 Message Block Chaining

As a solution to the problem of serialization caused by cipher block chaining, we propose a *message block chaining* mode of operation, whereby each block of plaintext is chained to the previous blocks of plaintext.

Let $m_0, m_1, m_2, \ldots$ be the original plaintext message, before chaining. Then we define $u_0, u_1, u_2, \ldots$ to be the chained message, after message block chaining, where

$u_0 = m_0$, $u_1 = u_0 \oplus m_1$, $u_2 = u_1 \oplus m_2$, ... (That is, u_0 is defined to be m_0 and u_i is defined to be $u_{i-1} \oplus m_i$, for $i > 0$.)

Like cipher block chaining, message block chaining is inherently sequential. However, message block chaining differs from cipher block chaining in that a chaining operation need not be combined with an enciphering operation in a single operation. Rather, message block chaining may be implemented as a preprocessing step performed prior to enciphering. Once the preprocessing step is complete, each chained message block u_i may be enciphered independently to form ciphertext block $c_i = E(u_i)$. Therefore, the chained message blocks may be enciphered in parallel.

Since the message block chaining operation involves only a bitwise exclusive-or operation, it requires very little time compared to the lengthy RSA enciphering operation. Consequently, performing the chaining operation sequentially has a negligible effect on the efficiency of parallel enciphering.

Message block chaining provides many of the desirable properties of cipher block chaining. Like cipher block chaining, duplicate portions of plaintext have a very low probability of being encoded into duplicate portions of ciphertext. Like cipher block chaining, each ciphertext block depends upon all previous blocks of the message, so that blocks of ciphertext cannot be cut and pasted to forge messages.

Message block chaining also allows individual blocks of ciphertext to be decoded without deciphering the entire ciphertext message. To decode ciphertext block c_i (where $i > 0$), we need only decipher ciphertext blocks c_i and c_{i-1} to get chained message blocks $u_i = D(c_i)$ and $u_{i-1} = D(c_{i-1})$. Plaintext block m_i corresponding to ciphertext block c_i is then simply

$$u_{i-1} \oplus u_i = u_{i-1} \oplus (u_{i-1} \oplus m_i)$$
$$= (u_{i-1} \oplus u_{i-1}) \oplus m_i$$
$$= 0 \oplus m_i$$
$$= m_i$$

Using message block chaining, $n + 1$ deciphering operations are required to decode a segment of n contiguous ciphertext blocks. Using cipher block chaining, only n deciphering operations are required to perform the same task. Therefore, when decoding only portions of a ciphertext, message block chaining requires slightly more computation than cipher block chaining.

As with cipher block chaining, message block chaining also allows the deciphering algorithm to recover from temporary transmission or enciphering errors. For example, suppose ciphertext block c_i contains an error. Only plaintext blocks $m_i = D(c_{i-1}) \oplus D(c_i)$ and $m_{i+1} = D(c_i) \oplus D(c_{i+1})$ depend on the value of c_i. Consequently, only those two blocks are affected by the error.

```
procedure encipher;
var e,M,mi,ci,ui,ulast:integer;
begin
    ReadKey(e, M);
    ulast:=0;
    while not eof do begin
        ReadBlock(mi);
        chain(ui, ulast, mi);
        ci:=modpower(ui, e, M);
        WriteBlock(ci)
    end
end;
```

Algorithm 8.1: RSA enciphering with message block chaining

```
procedure chain(var ui,ulast:integer; mi:integer);
begin
    ui:=xor(ulast, mi);
    ulast:=ui
end;
```

Algorithm 8.2: Chaining algorithm

8.5 Sequential Algorithms

Algorithm 8.1 defines the sequential procedure for RSA enciphering using message block chaining. Initially, enciphering key $\langle e, M \rangle$ is read. Subsequently, plaintext blocks are repeatedly read from an input file, chained, enciphered and written to an output file, until the end of the input file is reached.

The operation *ReadKey(e, M)* reads an RSA exponent e and modulus M. The operation *ReadBlock(mi)* reads plaintext block mi. The operation *WriteBlock(ci)* writes ciphertext block ci. The operation *modpower(ui, e, M)* computes the value of ui^e mod M. The value of *ulast* is the value of the last chained message block. Initially, *ulast* is 0.

The *chain* procedure implements message block chaining of two blocks, and is defined by algorithm 8.2. The *xor* function computes the integer value corresponding to the bitwise exclusive-or of two blocks. *ulast* is updated for the next chaining operation.

Algorithm 8.3 defines the sequential procedure for deciphering a message enciphered using the RSA cryptosystem with message block chaining. Deciphering key $\langle d, M \rangle$ is first read. Subsequently, ciphertext blocks are read from an input file, deciphered, unchained and written to an output file, until the end of the input file is reached.

Algorithm 8.4 shows the *unchain* procedure, which implements message block unchaining of two blocks. The plaintext message block is constructed from the bitwise exclusive-or of ui and *ulast*. *ulast* is updated for the next unchaining operation.

```
procedure decipher;
var d,M,ci,mi,ui,ulast:integer;
begin
    ReadKey(d, M);
    ulast:=0;
    while not eof do begin
        ReadBlock(ci);
        ui:=modpower(ci, d, M);
        unchain(mi, ulast, ui);
        WriteBlock(mi)
    end
end;
```

Algorithm 8.3: RSA deciphering with message block unchaining

```
procedure unchain(var mi,ulast:integer; ui:integer);
begin
    mi:=xor(ulast, ui);
    ulast:=ui
end;
```

Algorithm 8.4: Unchaining algorithm

8.6 Parallel Enciphering and Deciphering

RSA enciphering can be implemented in parallel using synchronized servers by simple substitution of several data types and procedures. The distributor and collector processes provide an interface for the server pipeline.

The channel protocol for the generic synchronized servers program is defined by the channel declaration

```
channel = [global(GlobalDataTypes), data(InputTypes),
    result(ResultTypes), complete, eos];
```

For RSA enciphering, the input data is a plaintext block, and the output data is a ciphertext block. As a result, we substitute the data type *integer* for both *InputTypes* and *ResultTypes*. The global data needed by all servers is the enciphering key $\langle e, M \rangle$. Therefore, we replace *GlobalDataTypes* with two integers. As a result, the channel protocol for the RSA enciphering program is defined by

```
channel = [global(integer,integer), data(integer),
    result(integer), complete, eos];
```

The distributor process reads blocks of plaintext from the input file, performs message block chaining, and sends the blocks into the server pipeline for enciphering. As we mentioned earlier, message block chaining requires very little execution time. Therefore, the chaining operation is performed sequentially by the distributor.

The distributor is defined by algorithm 8.5. The generic program uses two termination signals (*complete* and *eos*) to allow repeated use of the server pipeline. (We

```
procedure distributor(out:channel);
var e,M,mi,ui,ulast:integer;
begin
    ReadKey(e, M);
    out!global(e, M);
    ulast:=0;
    while not eof do begin
        ReadBlock(mi);
        chain(ui, ulast, mi)
        out!data(ui);
    end;
    out!complete;
    out!eos
end;
```

Algorithm 8.5: The distribution procedure for enciphering

```
procedure collector(inp:channel);
var ci:integer;
    more:boolean;
begin
    more:=true;
    while more do
        poll
            inp?result(ci) -> WriteBlock(ci) |
            inp?complete -> more:=false
        end;
    inp?eos
end;
```

Algorithm 8.6: The collection procedure for enciphering

shall not make use of this feature in our program.)

The collector process receives blocks of ciphertext from the server pipeline and then writes them to the output file. The procedure for the collector is defined by algorithm 8.6.

A generic server process is defined in chapter 7. Substituting for the generic data types as described above, renaming some variables to remain consistent with our sequential algorithms, and substituting the enciphering operation as the local computation procedure, we get the RSA enciphering server procedure defined by algorithm 8.7. The procedures *broadcast*, *ShiftIn*, and *ShiftOut* are defined in chapter 7 and require only generic data type substitution to be specialized for RSA enciphering.

The operation *broadcast(e, M, continue, id, inp, out)* implements a broadcast phase, during which the enciphering exponent and modulus are distributed to each of the server processes. The servers are capable of enciphering multiple messages. (We have defined a distributor and collector that encipher just a single message, however). The variable *continue* is updated to indicate whether the server should terminate.

The operation *ShiftIn(ui, more, id, inp, out)* implements a shift-in phase, during

```
procedure server(id:integer; inp,out:channel);
var e,M,ui,ci:integer;
    continue,more:boolean;
begin
    broadcast(e, M, continue, id, inp, out);
    while continue do begin
        more:=true;
        while more do begin
            ShiftIn(ui, more, id, inp, out);
            if more then begin
                ci:=modpower(ui, e, M);
                ShiftOut(ci, more, id, inp, out)
            end
        end;
        broadcast(e, M, continue, id, inp, out);
    end
end;
```

Algorithm 8.7: The RSA enciphering server procedure

which distinct chained message blocks are distributed to each server, if possible. The variable ui holds the current message block (if any) that is assigned to the given server. If there are fewer message blocks than servers, some of the servers will be idle. The variable *more* is updated to indicate whether or not the server is idle.

The operation *ShiftOut(ci, more, id, inp, out)* implements a shift-out phase, during which the ciphertext blocks are output to the collector. The variable ci holds the current ciphertext block computed by the given server. Boolean variable *more* is updated to indicate whether or not a completion signal (*complete*) has been received.

Due to the symmetry of RSA enciphering and deciphering, the procedure of algorithm 8.7 may also be used for RSA deciphering. (We have chosen the variable names to correspond to enciphering. Replacing e by d and switching the variable names ui and ci would correspond to deciphering.)

In order to specialize the program for deciphering, we need only modify the distributor and collector processes, so that they perform the necessary message block unchaining. The necessary changes are shown in algorithms 8.8 and 8.9.

```
procedure distributor(out:channel);
var d,M,ci:integer;
begin
    ReadKey(d, M);
    out!global(d, M);
    while not eof do begin
        ReadBlock(ci);
        out!data(ci);
    end;
    out!complete;
    out!eos
end;
```

Algorithm 8.8: The distribution procedure for deciphering

```
procedure collector(inp:channel);
var ui,mi,ulast:integer;
   more:boolean;
begin
   ulast:=0;
   more:=true;
   while more do
      poll
         inp?result(ui) ->
            unchain(mi, ulast, ui);
            WriteBlock(mi) |
         inp?complete ->
            more:=false
      end;
   inp?eos
end;
```

Algorithm 8.9: The collection procedure for deciphering

Together with the remainder of the of the generic program defined in chapter 7, this completes the distributed programs for RSA enciphering and deciphering.

8.7 Implementation Concerns

Depending upon the system used for implementation of the distributed programs for RSA enciphering and deciphering, performance may be improved by the addition of buffering. If the filing system is relatively slow, a buffer can be used to allow plaintext blocks to be read, and ciphertext blocks to be written, while the server pipeline is performing enciphering or deciphering. Furthermore, buffer processes placed between the servers processes in the pipeline may improve performance on some systems. We have used both of these buffering schemes for our implementation on a Meiko Computing Surface [Meiko 88].

The integer arithmetic in the algorithms above must be replaced with *multiple-length arithmetic*, since the integers involved are far larger than the word-size on most computers. The programming of the multiple-length operations is straightforward, with the notable exception of multiple-length division [Brinch Hansen 92d].

8.8 Performance

We reprogrammed the distributed algorithms for RSA enciphering and deciphering using occam 2 on a Meiko Computing Surface with T800 transputers. Table 8.1 shows sample throughput, speedup and efficiency figures for distributed RSA enciphering with a modulus and enciphering exponent of 640 bits each. Enciphering a single 640-bit block of data requires approximately ten seconds on a single transputer. The performance of deciphering is comparable.

To avoid limiting the performance of our program with a slow filing system, we tested the program using internally-stored data. These figures are shown by table 8.2.

Processors	Throughput (bit/s)	Speedup	Efficiency
1	61.9	1.00	1.00
5	309	4.99	1.00
10	610	9.85	0.99
20	1220	19.7	0.99
40	2420	39.1	0.98

Table 8.1: *Throughput, speedup and efficiency with 640-bit exponent and modulus*

Processors	Throughput (bit/s)	Speedup	Efficiency
1	61.9	1.00	1.00
5	310	5.00	1.00
10	619	10.0	1.00
20	1240	20.0	1.00
40	2470	40.0	1.00

Table 8.2: *Performance figures without filing system overhead*

Clearly, the distributed program is capable of achieving nearly-perfect speed-up. Performance tests indicate that the serial overhead for chaining or unchaining is negligible, reducing the throughput on forty processor by less than two tenths of a percent.

The speed of RSA enciphering may be increased by using a small enciphering exponent, provided that the exponent is not too small [Hastad 85, Jung 87]. Table 8.3 shows sample throughput, speedup and efficiency figures for distributed RSA enciphering with a modulus of 640 bits and a small enciphering exponent of 32 bits.

When a small exponent is used, the time required to encipher a single block of data on a single transputer drops to just 435 milliseconds. Still, the program is capable of achieving nearly-perfect speed-up, running with 98 percent efficiency on forty processors, for the sample shown.

Since our software implementation is rather slow compared to special-purpose hardware implementations, we also gather performance figures using a 640 bit modulus with an enciphering exponent of 3. Though this exponent is much too small to be safe for actual use, this allows us to simulate the performance of our parallel algorithm when the enciphering time for a single block is just 23.5 milliseconds.

The results are shown by table 8.4. While the computation time for enciphering a single block has decreased by a factor of over four hundred (compared to a 640 bit exponent), the communication overhead remains constant. Even so, the distributed program provides good performance, with an efficiency of 0.78 on forty processors.

Since a p-processor implementation of the distributed program processes p blocks of a file at a time, there will be some idle processor time if the number of blocks in a file to be enciphered or deciphered is not a multiple of p. For large files, this idle time will be a small fraction of the overall processing time. However, for small files, the overhead could be considerable. If efficiency is of primary concern, a moderate number of processors should be used when processing small files.

Processors	Throughput (Kbit/s)	Speedup	Efficiency
1	1.47	1.00	1.00
5	7.34	4.98	1.00
10	14.6	9.93	0.99
20	29.2	19.8	0.99
40	58.0	39.3	0.98

Table 8.3: Throughput, speedup and efficiency with 32-bit exponent

Processors	Throughput (Kbit/s)	Speedup	Efficiency
1	27.2	1.00	1.00
5	124	4.54	0.91
10	247	9.07	0.91
20	464	17.0	0.85
40	847	31.1	0.78

Table 8.4: Throughput, speedup and efficiency with exponent of 3

Chapter 9

Generating Deterministically Certified Primes

"From principles is derived probability, but truth or certainty is obtained only from facts."

Nathaniel Hawthorne

9.1 Introduction

The technique conventionally used to certify prime numbers is a probabilistic algorithm known as the Miller-Rabin test [Miller 76, Rabin 80]. This algorithm detects primes with a vanishing probability of error. While nonprobabilistic certification algorithms do exist [Adleman 87, Goldwasser 86], they are impractically slow to use [Beauchemin 86].

The problem of generating a prime is notably different from the problem of testing an arbitrary number for primality, however. When testing an arbitrary number, we cannot presume that the number has any special characteristics. For the problem of generating a prime, however, we can choose to generate only numbers that have special characteristics. Correspondingly, we can apply a primality test that requires those special characteristics.

This chapter describes a prime generation algorithm that produces deterministically certified primes. We then develop a distributed implementation of the algorithm from a generic program for the synchronized servers paradigm [Chapter 7] and from a generic program for the competing servers paradigm [Chapter 4].

9.2 The Primality Test

Shawe-Taylor [86] and Maurer [89] describe algorithms to generate certified primes based on Pocklington's Theorem [Bressoud 89]. To certify an integer n as prime, the certification algorithm requires a partial factorization of $n-1$. Correspondingly, these algorithms generate n so that a suitable partial factorization of $n-1$ is known. These algorithms, however, are not amenable to a parallel implementation.

We shall base our prime generation algorithm on a related theorem due to Edouard Lucas [Pomerance 87, Pratt 75, Ribenboim 91]. To certify an integer n as prime, the certification algorithm requires a complete factorization of $n-1$.

Lucas' Theorem: Let n be a natural number. If there exists an integer b such that $1 < b < n$ and

$$b^{n-1} \equiv 1 \pmod{n},$$

and for every prime factor q of $n-1$,

$$b^{(n-1)/q} \not\equiv 1 \pmod{n},$$

then n is prime.

If we have an integer n that we believe to be prime, and we know the factors of $n-1$, we can use Lucas' Theorem as a primality test. To prove that n is prime, we simply need to find a base b that satisfies the requirements of the theorem.

It turns out that if n is prime, an arbitrarily selected b is very likely to satisfy Lucas' criteria [Shawe-Taylor 86]. To develop a simple algorithm, then, we can arbitrarily fix a value for b, say 2, and test all candidates for primality using that base. Our experiments indicate that this works well in practice.

A general-purpose prime certification algorithm based on this theorem would be impractical, since the time required to factor $n-1$, for arbitrary n, is prohibitive. If a factorization of $n-1$ is known, however, the certification test is quite fast. To be practical, then, our prime generation algorithm will have to generate n so that the factorization of $n-1$ is known.

9.3 Generating Suitable Prime Candidates

We propose an algorithm whereby a candidate for primality testing, n, is chosen so that $n-1$ has only small prime factors (for example, uniformly drawn from among the ninety-five primes less than five hundred). While this substantially restricts the pool of numbers from which we draw candidates for testing, the pool is still quite large.

Consider generating a one hundred digit number. Without restrictions, we have a pool of approximately 10^{100} numbers from which to draw. We can select numbers with small prime factors only, by randomly selecting small prime factors until the product has one hundred digits.

If we choose from among the ninety-five primes less than five hundred, then we will need at least thirty-seven factors to form the product, since all of the factors are less than 500 and $500^{37} < 10^{100}$. As a result, we select a number from a pool of at least 95^{37}, or approximately 10^{73} numbers. The number of primes less than n is asymptotic to $n/(\ln n)$ [Bressoud 89, Riesel 85]. Therefore, of the pool of approximately 10^{73} numbers, we expect that more than 10^{70} will be prime.

This approach is quite different from the conventional approach for randomly generating factored numbers. Bach [84, 88] developed an algorithm to generate uniformly distributed random numbers with known factorization. Maurer [89] used this idea to generate primes with nearly-uniform distribution.

For the purposes of constructing RSA keys, however, it is not necessary to have a nearly-uniform distribution of primes. It is only necessary that the primes used do not make an RSA modulus susceptible to some special-purpose factoring attack, or to some other special-purpose cryptanalytic attack.

There is a well-known algorithm to quickly factor a product with prime factor p such that $p-1$ has only small prime factors [Pollard 74]. Therefore, it is not secure to use the primes of the special form we have described as factors of an RSA modulus. However, we can use these special primes to construct *strong primes* for RSA moduli [Gordon 84, Chapter 5].

In chapter 5, we discussed the construction of strong primes in detail. In this chapter, we shall concern ourselves primarily with the problem of generating primes of the special form described above. Later we will briefly discuss the problem of constructing strong primes for RSA moduli using these special primes.

9.4 Sequential Algorithm

Before presenting the sequential algorithm to generate primes, we introduce some definitions used in our algorithms. Constant *numprimes* defines the number of small prime factors to be used in generating primes. The array type

```
FactorArray=array [1..numprimes] of boolean;
```

represents the prime factors of a number. Lucas' Theorem requires that we know the prime factors of $n-1$, but it does not require that we know the powers of those prime factors. (For example, it is important to know whether 2 is a prime factor, but it is not important to know whether 2^2 or 2^3 is a factor.) As a result, we may use an array of booleans to indicate which of the small primes are factors of $n-1$.

Algorithm 9.1 defines a sequential procedure for generating prime numbers certified using Lucas' Theorem. *bits* is the number of bits in the generated number. n is a variable parameter to which the generated prime is assigned.

The algorithm repeatedly generates a random value of $n-1$, with a known factorization, until an n is found that satisfies Lucas' Theorem for $b=2$. The operation *modpower(2, n-1, n)* computes $2^{n-1} \bmod n$.

Since approximately one in every $\ln n$ integers close to n is prime, there is approximately a $1/\ln n$ probability that a randomly selected integer n is prime. Equivalently, the probability that a randomly selected integer n is composite is approximately $(\ln n - 1)/\ln n$.

```
procedure GeneratePrime(bits:integer; var n:integer);
var factor:FactorArray;
    prime:boolean;
begin
    prime:=false;
    repeat
        RandomProduct(bits, n, factor);
        n:=n+1;
        if modpower(2, n-1, n)=1 then certify(n, factor, prime)
    until prime
end;
```

Algorithm 9.1: Generating primes sequentially

For $n \approx 10^{100}$, $\ln n \approx 230$ so the probability that a randomly selected number close to 10^{100} is composite is $p_1 \approx 0.99565$. The probability that m randomly selected numbers close to 10^{100} will all be composite is $p_m \approx 0.99565^m$. Therefore, as $m \to \infty$, $p_m \to 0$. For example, $p_{1000} \approx 0.013$ and $p_{2000} \approx 0.00016$.

This means that there is a negligible probability that algorithm 9.1 will fail to find a prime. We can safely assume that algorithm 9.1 always terminates.

The *certify* procedure is algorithm 9.2. n is the number to be certified. *factor* is an array representing the factors of $n-1$. *prime* is a variable parameter used to indicate whether or not n was proven to be prime.

Prime candidate n is known to satisfy the first part of Lucas' Theorem, for a base of 2. That is, $2^{n-1} \bmod n = 1$. The certification algorithm tests whether n satisfies the second part of Lucas' Theorem: $b^{(n-1)/q} \not\equiv 1 \pmod{n}$ for every prime factor q of $n-1$. Since almost every number that satisfies the first part of Lucas' Theorem is prime [Jung 87, Rivest 90], we do not worry about terminating the certification procedure early, in the case where n is not proven to be prime.

Algorithm 9.3 is the procedure for generating a random number with known prime factors. *bits* is the size of the generated number. n is a variable parameter to which the generated number is assigned. *factor* is a variable parameter to which the factors of $n-1$ are assigned. *primes* is a previously-defined array containing the small primes to be used as factors.

```
procedure certify(n:integer; factor:FactorArray;
    var prime:boolean);
var i,f:integer;
begin
    prime:=true;
    for i:=1 to numprimes do
        if factor[i] then begin
            f:=(n-1)/primes[i];
            prime:=prime and (modpower(2, f, n)<>1)
        end
end;
```

Algorithm 9.2: Certifying a prime using Lucas' Theorem

```
procedure RandomProduct(bits:integer; var n:integer;
    var factor:FactorArray);
var i:integer;
begin
    n:=2;
    factor[1]:=true;
    for i:=2 to numprimes do factor[i]:=false;
    while size(n)<bits do begin
        i:=random(1, numprimes);
        if size(n*primes[i])<=bits then begin
            n:=n*primes[i];
            factor[i]:=true
        end
    end
end;
```

Algorithm 9.3: Generating a factored random number

The operation *size(x)* computes the number of bits used to represent the integer, *x*. The operation *random(1, numprimes)* selects a random number from 1 to *numprimes*.

The number being generated is initialized to have a factor of 2, since $n-1$ must be even, in order for *n* to be prime (for $n > 2$). The array of factors is initialized to reflect this. Subsequently, small prime factors for the number are randomly generated, until the random number has the desired number of bits.

Any factor that would make the generated number too large is rejected. Since multiplying any integer *n* by 2 increases *size(n)* by 1, algorithm 9.3 is always able to generate a number of exactly the specified size.

9.5 Parallel Algorithm

As in chapter 5, we separate the process of generating primes into two parts, for the purpose of parallelizing algorithm 9.1. The first part of the algorithm is the generation of prime candidates, where a prime candidate is a number that satisfies the first part of Lucas' Theorem (and therefore, is very likely to be prime). The second part is the certification of prime candidates.

We shall develop a distributed algorithm for prime candidate generation based upon the generic program for competing servers described in chapter 4. The resulting algorithm will be very similar to the distributed candidate generation algorithm in chapter 5.

Once a prime candidate is found, we must certify that it is prime. We shall use the generic program for synchronized servers described in chapter 7 to develop a distributed algorithm to perform prime certification based upon Lucas' Theorem.

9.6 Parallel Candidate Generation

The generic program for competing servers uses a channel protocol defined by the type declaration

```
channel = [start(SearchDataTypes), result(ResultTypes),
    check, continue, complete, eos];
```

To specialize the generic program to solve a particular problem, we must substitute for *SearchDataTypes* and *ResultTypes*.

For the process of generating prime candidates, we start with a value indicating the desired size, in bits, for the prime candidate. Therefore, we will substitute the type *integer* for *SearchDataTypes*. Once a suitable prime candidate n is found, it is returned along with the factorization of $n-1$. We, therefore, substitute *integer* and *FactorArray* for *ResultTypes*. Finally, to avoid a later conflict when we combine candidate generation and certification, we rename *result* to *candidate*. This makes the channel protocol for our program,

```
channel = [start(integer), candidate(integer,FactorArray),
    check, continue, complete, eos];
```

The heart of the competing servers program is a sequential search executed by each of the server processes. The search is periodically stopped to check whether the search should be terminated, as a result of some other server having found a solution. Algorithm 9.4 is a modified version of the generic procedure, specialized for the

```
procedure FindPrime(bits:integer; inp1,out1,out2:channel);
var n:integer;
    factor:FactorArray;
    more,found:boolean;
begin
    more:=true;
    while more do begin
        out1!check;
        poll
            inp1?complete -> more:=false |
            inp1?continue ->
                RandomProduct(bits, n, factor);
                n:=n+1;
                found:=(modpower(2, n-1, n)=1);
                if found then begin
                    out2!candidate(n, factor);
                    out1!complete;
                    more:=false
                end
        end
    end;
    out2!complete
end;
```

Algorithm 9.4: *Search procedure modified for prime candidate generation*

generation of prime candidates to be certified using Lucas' Theorem. The desired size of the generated prime is passed as a parameter.

Each step in the search consists of generating a random n such that the factorization of $n-1$ is known, and testing n to see if it satisfies the requirement that $2^{n-1} \bmod n = 1$. Each server process uses a local random number generator initialized with a unique seed value. Once a prime candidate is found, it is returned to the master process to undergo certification.

The parameters *inp1* and *out1* are channels connecting the searcher process to a local *reader* process. During the search for factors, each searcher periodically communicates with its local reader. The searcher sends a *check* signal to the reader. The reader responds with either a *complete* signal or a *continue* signal. A *complete* signal indicates that the search should be terminated. A *continue* signal indicates that the search should be continued [Chapter 4].

out2 is a channel connecting the *searcher* process to a local *writer* process. If the searcher finds a factor, it outputs that factor to its local writer.

As discussed in chapter 5, the process of generating prime candidates can be made more efficient by eliminating numbers that have small prime factors. In addition, we can speed up the generation of factored random numbers by starting with a random initial product (slightly smaller than the desired size), and merely extending that product, rather than generating a whole new product at each iteration.

```
procedure FindPrime(bits, x:integer; inp1,out1,out2:channel);
var n:integer;
    factor:FactorArray;
    more,found:boolean;
begin
    more:=true;
    while more do begin
        out1!check;
        poll
            inp1?complete -> more:=false |
            inp1?continue ->
                n:=x
                ExtendProduct(bits, n, factor);
                n:=n+1;
                found:=false;
                if not SmallFactor(n) then
                    found:=(modpower(2, n-1, n)=1);
                if found then begin
                    out2!candidate(n, factor);
                    out1!complete;
                    more:=false
                end
        end
    end;
    out2!complete
end;
```

Algorithm 9.5: Improved search procedure for prime candidate generation

```
procedure ExtendProduct(bits:integer; var n:integer;
    var factor:FactorArray);
var i:integer;
begin
    for i:=1 to numprimes do factor[i]:=false;
    while size(n)<bits do begin
        i:=random(1, numprimes);
        if size(n*primes[i])<=bits then begin
            n:=n*primes[i];
            factor[i]:=true
        end
    end
end;
```

Algorithm 9.6: Extending a factored random number

These small changes are included in algorithm 9.5. The search data carried by a *start* message now consists of two integers: the desired number of bits, and the initial product. The parameters of the search procedure include the initial product.

Algorithm 9.6 is the procedure for randomly extending an initial product. It is identical to algorithm 9.3, except that n is not initialized since it contains the initial product. The factors of the initial product are retained by the master process, so *factor* represents only those factors used to extend the initial product.

The *master* process executes algorithm 9.7. The master first generates the random initial product x. Constant *bitsLeft* denotes the number of bits to be left for the server processes to generate. The master then initiates a search by broadcasting a *start* message carrying the desired candidate size and the initial product. Since the server processes do not need the factors of the initial product, the master does not broadcast them.

To guarantee termination, *bitsLeft* should be large enough to create a large pool of numbers. For example, if *bitsLeft* = 32 and we are using prime factors less 500, at least three factors will be necessary to extend the initial product (since $500^3 < 2^{32}$). Therefore, the servers select from a pool of at least $95^3 \approx 850,000$ numbers. If we

```
procedure master(bits:integer; var n:integer;
    var factor:FactorArray; inp,out:channel);
var x,i:integer;
    f:FactorArray;
begin
    RandomProduct(bits-bitsLeft, x, f);
    out!start(bits, x);
    inp?candidate(n, factor);
    for i:=1 to numprimes do factor[i]:=factor[i] or f[i];
    out!complete;
    inp?complete;
    out!eos
end;
```

Algorithm 9.7: Master procedure for prime candidate generation

are generating primes of approximately 100 decimal digits, the probability of failing to find a prime within this pool is less than $p_{850,000} < 10^{-1600}$, which is negligible.

After initiating a search, the master awaits a *result* message carrying prime candidate n and the factors used to extend the initial product. The returned list of factors is combined with the list of factors for the initial product to form a complete list of factors for $n - 1$.

The remaining communications are used to complete a search and to terminate the server processes. The details of search completion and process termination are discussed in chapter 4.

Together with the remainder of the competing servers program shell described in chapter 4, this defines a complete distributed algorithm to generate prime candidates. The distributed certification algorithm is described below.

9.7 Parallel Certification

The channel protocol for the generic synchronized servers program is defined by the channel declaration

```
channel = [global(GlobalDataTypes), data(InputTypes),
   result(ResultTypes), complete, eos];
```

To specialize the generic program, we must substitute for *GlobalDataTypes*, *Input-Types* and *ResultTypes*.

The prime certification algorithm based on Lucas' Theorem consists of testing the primality of prime candidate n by testing that $b^{(n-1)/q} \not\equiv 1 \pmod{n}$ for every prime factor q of $n - 1$. Every server process requires n, so *GlobalDataTypes* is replaced with *integer*. The factors of $n - 1$ are distributed among the servers, so *InputTypes* is also replaced by *integer*. The result of each server's computation is a boolean indicating whether or not the test was passed. Therefore, we substitute *boolean* for *ResultTypes*.

The distributor process for prime certification is defined by algorithm 9.8. The process first broadcasts prime candidate n to all of the server processes. Subsequently, the distributor outputs the prime factors of $n - 1$ into the server pipeline.

```
procedure distributor(n:integer; factor:FactorArray;
   out:channel);
var i:integer;
begin
   out!global(n);
   for i:=1 to numprimes do
      if factor[i] then out!data(primes[i]);
   out!complete;
   out!eos
end;
```

Algorithm 9.8: *Distribution procedure*

```
procedure certifier(id:integer; inp,out:channel);
var n,q:integer;
    pass,continue,more:boolean;
begin
    broadcast(n, continue, id, inp, out);
    while continue do begin
        more:=true;
        while more do begin
            ShiftIn(q, more, id, inp, out);
            if more then begin
                pass:=(modpower(2, (n-1) div q, n)<>1);
                ShiftOut(pass, more, id, inp, out)
            end
        end;
        broadcast(n, continue, id, inp, out);
    end
end;
```

Algorithm 9.9: The prime certification server procedure

The factors are distributed among the server processes as evenly as possible, to balance the work load.

Algorithm 9.9 is the generic server procedure modified for prime certification. Each certifier tests a distinct prime factor q of $n-1$ to determine whether $b^{(n-1)/q} \not\equiv 1 \pmod{n}$. Boolean variable *pass* is used to indicate whether or not the primality test was passed for factor q. Depending upon the number of factors to be tested, each server may test a number of different factors.

We have renamed the server procedure *certifier*, to avoid a name conflict when we combine candidate generation with certification. The *broadcast*, *ShiftIn* and *ShiftOut* procedures are defined in chapter 7 and require only generic data type substitution.

The operation *broadcast(n, continue, id, inp, out)* implements a broadcast phase, during which prime candidate n is distributed to each of the server processes. The servers are capable of testing multiple candidates. The variable *continue* is used to indicate when the server should terminate.

The operation *ShiftIn(q, more, id, inp, out)* implements a shift-in phase, during

```
procedure collector(var prime:boolean; inp:channel);
var more,pass:boolean;
begin
    prime:=true;
    more:=true;
    while more do
        poll
            inp?result(pass) -> prime:=prime and pass |
            inp?complete -> more:=false
        end;
    inp?eos
end;
```

Algorithm 9.10: Collection procedure

which distinct factors of $n - 1$ are distributed to each server, if possible. If there are fewer factors than servers, some of the servers will be idle. The variable *more* is used to indicate whether or not the server is idle.

The operation *ShiftOut(pass, more, id, inp, out)* implements a shift-out phase, during which the results of the individual primality tests are output to the collector. Boolean variable *more* is used to indicate whether or not a completion signal has been received.

The collection process executes algorithm 9.10. The tested number is proven prime only if all of the individual tests, performed by the servers, are passed.

9.8 Combining Generation and Certification

We must combine prime candidate generation and certification to construct a complete distributed algorithm to generate primes. Figure 9.1 shows the topology of the complete distributed program. The competing servers master process is combined with the synchronized servers distributor. The collector process outputs its results directly to that master process.

For convenience, we combine the two channel protocols into a single protocol defined by the type declaration:

```
channel = [start(integer,integer),
    candidate(integer,FactorArray), check, continue,
    complete, eos, global(integer), result(boolean)];
```

The combined *master* process executes the procedure defined by algorithm 9.11. As before, the master first generates the random initial product.

The master then initiates a search and waits for a prime candidate and the corresponding factors. After the returned list of factors is combined with the list of factors for the initial product, the candidate is submitted for certification. If necessary, the search and certification is repeated until a prime is certified.

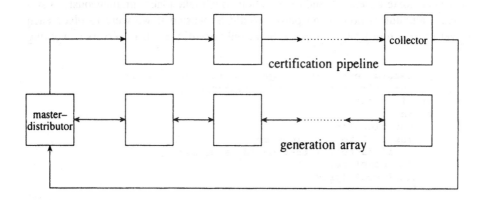

Figure 9.1: *Combined topology*

```
procedure master(bits:integer; var n:integer;
    inp1,out1,inp2,out2:channel);
var x,i:integer;
    factor,f:FactorArray;
    prime:boolean;
begin
    RandomProduct(bits-bitsLeft, x, f);
    repeat
        out2!start(bits, x);
        inp2?candidate(n, factor);
        for i:=1 to numprimes do factor[i]:=factor[i] or f[i];
        out2!complete;
        certify(n, factor, prime, inp1, out1);
        inp?complete
    until prime;
    out2!eos;
    out1!eos
end;
```

Algorithm 9.11: Complete master procedure for prime generation

The distributor process is integrated into the master process as the certification operation. Algorithm 9.12 is the certification procedure. Prime candidate n and the corresponding factors of $n-1$ are output, as before. A final result indicating whether or not the candidate is prime is input from the collector process.

The collection procedure is modified to handle multiple certifications, and to output its result to the master. Since the collector receives all of the results before returning a single result, it acts as a buffer and there is no chance of deadlock. Algorithm 9.13 defines the modified collector.

The parallel network combining the master process with the collector and the two sets of servers is defined by algorithm 9.14. (The *generator* statement corresponds to the *server* statement in chapter 4.)

To generate a certified prime, we first generate prime candidate n via a parallel search, and then certify that candidate by testing the factors of $n-1$ in parallel. Since prime candidate generation and certification do not take place simultaneously, a significant amount of processing power would be wasted if we were to place each server process on its own processor. Since only one of the sets of servers is working

```
procedure certify(n:integer; factor:FactorArray;
    var prime:boolean; inp,out:channel);
var i:integer;
begin
    out!global(n);
    for i:=1 to numprimes do
        if factor[i] then out!data(primes[i]);
    out!complete;
    inp?result(prime)
end;
```

Algorithm 9.12: Certification procedure

```
procedure collector(inp,out:channel);
var more,pass,prime:boolean;
begin
    prime:=true;
    more:=true;
    while more do
        poll
            inp?result(pass) -> prime:=prime and pass |
            inp?complete ->
                out!result(prime);
                prime:=true |
            inp?eos -> more:=false
        end
end;
```

Algorithm 9.13: Modified collection procedure

```
procedure GeneratePrime(bits:integer; var x:integer);
type net=array [0..p] of channel;
var a,b,c:net;
    d:channel;
    i:integer;
begin
    parbegin
        master(bits, x, d, a[p], b[0], c[0])|
        collector(a[0], d)
        parfor i:=1 to p do
            parbegin
                certifier(i, a[i], a[i-1])|
                generator(i, c[i-1], c[i], b[i], b[i-1])
            end
    end
end;
```

Algorithm 9.14: Arranging the overall computation

at any one time, we can achieve much better efficiency by placing one candidate generating server process and one certification server process on each processor.

To achieve this improvement, we need only insure that the innermost *parbegin* of algorithm 9.14 is executed by a single processor. We also improve efficiency by executing the master and the collector on a single processor.

9.9 Performance

We reprogrammed the distributed algorithm for prime generation using occam 2 on a Meiko Computing Surface with T800 transputers. To test our parallel algorithm, we generated certified primes with 256, 320 and 512 bits (corresponding to approximately 75, 100, and 150 decimal digits, respectively).

The running time for a sequential prime number generator is not constant, even for generated primes of constant size. This is because the primes are not distributed uniformly and, depending upon our luck, it may take some time before we discover one.

Processors	Running Time (s)	Speedup	Efficiency
1	24	1.0	1.00
5	6.7	3.6	0.72
10	5.0	4.8	0.48
20	3.9	6.2	0.31
40	3.8	6.3	0.16

Table 9.1: *Average running times for 256-bit prime generation*

Processors	Running Time (s)	Speedup	Efficiency
1	55	1.0	1.00
5	15	3.6	0.72
10	9.9	5.6	0.56
20	7.7	7.1	0.36
40	6.5	8.5	0.21

Table 9.2: *Average running times for 320-bit prime generation*

In addition, the non-deterministic nature of our parallel candidate generation algorithm will cause the particular prime generated (and consequently, the running time of the algorithm) to vary with the number of processors used.

Since approximately one in every $\ln n$ numbers around n is prime, testing should be averaged over $O(\ln n)$ trials, ideally. Tables 9.1, 9.2 and 9.3 show the average running times for the generation of 256, 320 and 512 bit primes, respectively. The running times are averaged over two hundred trials. In each case, the parallel run time on forty processors is an order of magnitude smaller than the sequential run time.

As the figures of tables 9.1, 9.2 and 9.3 indicate, the distributed program is inefficient when executed on a large number of processors. This is a result of two problems. First, the prime generation algorithm loses efficiency when executed on many processors [Chapter 5].

Furthermore, the prime generation in this program is less efficient than that of chapter 5. This is because a higher percentage of the random numbers generated do not have any small prime factors. (When $n-1$ is chosen to have small factors, we decrease the likelihood that n will contain small factors.) Therefore, the slow Miller-Rabin test is applied more frequently using algorithm 9.4 than the corresponding algorithm in chapter 5.

Secondly, during the certification phase, the parallelism is limited by the number of factors to be tested. When the number of factors is not evenly divisible by the number of server processes, some server process idle time will result. This reduces processor efficiency.

For example, in the case of 320-bit primes, there are typically just over forty factors to be tested. When running on forty processors, this means that the factors are tested by the pipeline in two sets. The first forty factors are tested first. Subsequently, any remaining factors are tested. Since there are only a few remaining factors, most

Processors	Running Time (s)	Speedup	Efficiency
1	340	1.0	1.00
5	74	4.6	0.92
10	48	7.1	0.71
20	35	9.7	0.49
40	27	13	0.33

Table 9.3: *Average running times for 512-bit prime generation*

of the servers are idle during the testing of the second set, and the efficiency is poor. The same phenomenon occurs for lower numbers of processors, though with fewer processors, the percentage of idle time is reduced.

The sequential performance of the algorithm is comparable to that of the conventional Miller-Rabin algorithm [Chapter 5].

9.10 Generating Strong Primes

As we discussed earlier, the primes generated using this algorithm should not be used as factors of an RSA modulus, because of the existence of the *Pollard $p-1$ method* of factoring [Pollard 74, Chapters 5, 6]. However, primes generated by this algorithm can be used to construct strong primes for RSA moduli.

Prime p is a strong prime when

$$p \equiv 1 \pmod{r},$$
$$p \equiv s - 1 \pmod{s}, \tag{9.1}$$
$$r \equiv 1 \pmod{t},$$

where r, s and t are large primes [Chapter 5].

In general, we cannot generate $p \equiv s - 1 \pmod{s}$ for given prime s such that we know the factorization of $p-1$. Therefore, we cannot use our algorithm based on Lucas' Theorem to certify such a p. We propose two ways to remedy this problem.

A prime p with the property $p \equiv s - 1 \pmod{s}$ for some large prime s may be used to form a modulus that is secure against factorization by the *Williams $p+1$ method* [Williams 82, Chapter 5]. Though our algorithm cannot construct a prime p for which s is known, most primes do satisfy this requirement. Therefore, one approach to generating strong primes using our algorithm is to construct primes that satisfy the other two requirements of (9.1). Subsequently, we can insure that a modulus constructed from those primes is secure by testing it with a $p+1$ factoring attack [Chapter 6].

A second way to generate strong primes with our algorithm is to use a hybrid approach. The prime generation algorithm based upon Pocklington's Theorem, that we mentioned earlier, is capable of generating strong primes proven to satisfy the requirements of a strong prime [Shawe-Taylor 86]. However, that algorithm is not amenable to a parallel implementation, because it generates primes recursively.

Generating a strong prime that satisfies all of the requirements of (9.1) requires the generation of arbitrary primes s and t, and the subsequent generation of special primes r and p. A hybrid approach would use our algorithm to generate simple primes s and t. After this is done, the algorithm based on Pocklington's Theorem would be used to generate special primes r and p. This would allow a substantial amount of work to be performed in parallel, while generating primes proven to be strong.

Appendix 1

A Proof of Lucas' Theorem

Lucas' Theorem: Let n be a natural number. If there exists an integer b such that $1 < b < n$ and

$$b^{n-1} \equiv 1 \pmod{n},$$

and for every prime factor q of $n - 1$,

$$b^{(n-1)/q} \not\equiv 1 \pmod{n},$$

then n is prime.

Our proof is patterned after [Ribenboim 91]. Recall that $\phi(n)$ is the Euler totient function, defined to be the number of positive integers less than n that are relatively prime to n [Niven 80]. Before we start, we need a definition and two lemmas.

Definition: The order of b modulo n is the smallest positive integer e such that $b^e \equiv 1 \pmod{n}$.

Lemma 1: If e is the order of b modulo n, and $b^m \equiv 1 \pmod{n}$, e divides m.

Proof: This is proven by contradiction. Assume that e does not divide m. There must exist two integers u and v such that $m = ue + v$, where $1 \le v < e$.

Since $b^m \equiv 1 \pmod{n}$, we have

$$1 \equiv b^{ue+v} \pmod{n} \qquad\qquad \textit{by definition of m}$$

$$\equiv \left(b^e\right)^u b^v \pmod{n}$$

$$\equiv 1^u b^v \pmod{n} \qquad\qquad \textit{by definition of e}$$

$$\equiv b^v \pmod{n}$$

By definition, e is the smallest positive integer such that $b^e \equiv 1 \pmod{n}$. However, since $v < e$ and $b^v \equiv 1 \pmod{n}$ we have a contradiction. So e divides m.

End Proof.

Lemma 2: If e is the order of b modulo n, and $b^m \not\equiv 1 \pmod{n}$, e does not divide m.

Proof: Again we prove by contradiction. Assume that e divides m. There must exist an integer u such that $m = ue$. Since $b^m \not\equiv 1 \pmod{n}$, we have

$$1 \not\equiv b^{ue} \pmod{n} \qquad\qquad \textit{by definition of m}$$

$$1 \not\equiv \left(b^e\right)^u \pmod{n}$$

$$1 \not\equiv 1^u \pmod{n} \qquad\qquad \textit{by definition of e}$$

$$1 \not\equiv 1 \pmod{n}$$

which is a contradiction. Therefore, e does not divide m.

End Proof.

Now we are ready to prove Lucas' Theorem.

Proof of Lucas' Theorem: n is prime if and only if $\phi(n) = n - 1$. Since $\phi(n) \leq n - 1$, n is prime if $n - 1$ divides $\phi(n)$.

Assume that $n - 1$ does not divide $\phi(n)$. For this to be true, $n - 1$ must have at least one factor that $\phi(n)$ does not have. More specifically, there must exist a prime, q, and an integer, $r \geq 1$, such that q^r divides $n - 1$ but not $\phi(n)$. Let u be an integer such that $n - 1 = uq^r$.

Now, let e be the order of b modulo n. By definition, e is the smallest positive integer such that $b^e \equiv 1 \pmod{n}$.

By the supposition of Lucas' Theorem we have

$$b^{n-1} \equiv 1 \pmod{n}.$$

Since e is the order of b modulo n, and $b^{n-1} \equiv 1 \pmod{n}$, e divides $n - 1$ (by Lemma 1).

Also by the supposition of the theorem we have

$$b^{(n-1)/q} \not\equiv 1 \ (\text{mod } n).$$

Since e is the order of b modulo n, and $b^{(n-1)/q} \not\equiv 1 \ (\text{mod } n)$, e does not divide $(n-1)/q$ (by Lemma 2).

Since e divides $n-1 = uq^r$ but does not divide $(n-1)/q = uq^{r-1}$, e must have a factor of q^r.

Now, by Euler's generalization of Fermat's Theorem [Niven 80],

$$b^{\phi(n)} \equiv 1 \ (\text{mod } n).$$

Since e is the order of b modulo n, and $b^{\phi(n)} \equiv 1 \ (\text{mod } n)$, e divides $\phi(n)$ (by Lemma 1).

Finally, since e has a factor of q^r and e divides $\phi(n)$, $\phi(n)$ must also have a factor of q^r. But this is a contradiction. Therefore, $n-1$ divides $\phi(n)$, and n is prime.

End Proof.

Appendix 2

Multiple-Length Arithmetic

This appendix includes the occam 2 implementation of multiple-length arithmetic that is common to the programs contained in the following five appendices. The algorithms are based on those developed by Per Brinch Hansen but have been modified to use low-level occam routines that support a base-2^{32} representation.

To avoid repetition, the code presented here has been removed from the programs presented in the following appendices. Also excluded from those programs is the file interface (written by Brinch Hansen), and the definitions of some simple terminal I/O routines.

Finally, separate compilation in occam 2 prevents independent units from sharing constants and types. Consequently, declarations of such constants and types must be repeated within each compilation unit. Generally, I have suppressed the repeated listing of these declarations by placing all but one of the declarations into hidden folds. Most notably, global constant and type definitions appear only at the beginning of each program.

```
[W]INT zero, one, two, maxlong:
INT64 seed:

PROC overflow()
  STOP
:

INT FUNCTION min(VAL INT x, y)
  INT f:
  VALOF
    IF
      x <= y
        f := x
      TRUE
        f := y
    RESULT f
:

INT FUNCTION length(VAL [W]INT x)
  -- length(xk...x1x0) = k+1
  INT i, j:
  VALOF
    SEQ
      i, j := w, 0
      WHILE i <> j
        IF
          x[i] <> 0
            j := i
          TRUE
            i := i - 1
    RESULT i + 1
:

INT FUNCTION shortint(VAL [W]INT x) IS
  -- shortint(x) = integer(x)
  x[0]:

PROC longint([W]INT x, VAL INT y)
  -- x := long(y)
  SEQ
    x[0] := y
    SEQ i = 1 FOR w
      x[i] := 0
:

BOOL FUNCTION less(VAL [W]INT x, y)
  INT i, j, borrow, temp:
  VALOF
    SEQ
      i, j := w, 0
      WHILE i <> j
        IF
          x[i] <> y[i]
            j := i
          TRUE
            i := i - 1
      borrow, temp := LONGDIFF(x[i], y[i], 0)
    RESULT (borrow = 1)
:

BOOL FUNCTION equal(VAL [W]INT x, y)
  INT i, j:
  VALOF
    SEQ
```

```
      i, j := w, 0
      WHILE i <> j
        IF
          x[i] <> y[i]
            j := i
          TRUE
            i := i - 1
    RESULT x[i] = y[i]
:

BOOL FUNCTION greater(VAL [W]INT x, y)
  INT i, j, borrow, temp:
  VALOF
    SEQ
      i, j := w, 0
      WHILE i <> j
        IF
          x[i] <> y[i]
            j := i
          TRUE
            i := i - 1
      borrow, temp := LONGDIFF(y[i], x[i], 0)
    RESULT (borrow = 1)
:

BOOL FUNCTION odd(VAL [W]INT x)
  INT dummy, remainder:
  VALOF
    SEQ
      dummy, remainder := LONGDIV(0, x[0], 2)
    RESULT (remainder = 1)
:

PROC product([W]INT x, y, VAL INT k)
  -- x := y * k
  INT carry, m:
  SEQ
    m := length(y)
    x := zero
    carry := 0
    SEQ i = 0 FOR m
      carry, x[i] := LONGPROD(y[i], k, carry)
    IF
      m <= w
        x[m] := carry
      carry <> 0
        overflow()
      TRUE
        SKIP
:

PROC quotient([W]INT x, y, VAL INT k)
  -- x := y div k
  INT carry, i:
  SEQ
    x := zero
    carry := 0
    i := length(y)-1
    WHILE i >= 0
      SEQ
        x[i], carry := LONGDIV(carry, y[i], k)
        i := i-1
:
```

```
PROC remainder([W]INT x, y, VAL INT k)
  -- x := y mod k
  INT carry, i, dummy:
  SEQ
    x := zero
    carry := 0
    i := length(y)-1
    WHILE i >= 0
      SEQ
        dummy, carry := LONGDIV(carry, y[i], k)
        i := i-1
    x[0] := carry
:

PROC sum([W]INT x, VAL INT k)
  -- x := x+k
  INT carry, i:
  SEQ
    carry, x[0] := LONGSUM(x[0], k, 0)
    i := 1
    WHILE (carry<>0) AND (i<w)
      SEQ
        carry, x[i] := LONGSUM(x[i], 0, carry)
        i := i+1
    IF
      (i=w) AND (carry<>0)
        overflow()
      TRUE
        SKIP
:


PROC increment([W]INT x)
  -- x := x + 1
  INT carry, i:
  IF
    less(x, maxlong)
      SEQ
        carry := 1
        i := 0
        WHILE carry <> 0
          SEQ
            carry, x[i] := LONGSUM(x[i], 0, carry)
            i := i+1
    TRUE
      overflow()
:

PROC decrement([W]INT x)
  -- x := x - 1
  INT borrow, i:
  IF
    greater(x, zero)
      SEQ
        borrow := 1
        i := 0
        WHILE borrow <> 0
          SEQ
            borrow, x[i] := LONGDIFF(x[i], 0, borrow)
            i := i+1
    TRUE
      overflow()
:
```

```
PROC double([W]INT x)
  -- x := 2*x
  [W]INT x0:
  SEQ
    x0 := x
    product(x, x0, 2)
:

PROC halve([W]INT x)
  -- x := x/2
  [W]INT x0:
  SEQ
    x0 := x
    quotient(x, x0, 2)
:

PROC add([W]INT x, y)
  -- x := x + y
  INT carry:
  SEQ
    carry := 0
    SEQ i = 0 FOR w
      carry, x[i] := LONGSUM(x[i], y[i], carry)
    IF
      carry <> 0
        overflow()
      TRUE
        SKIP
:

PROC subtract([W]INT x, y)
  -- x := x - y
  INT borrow:
  SEQ
    borrow := 0
    SEQ i = 0 FOR w
      borrow, x[i] := LONGDIFF(x[i], y[i], borrow)
    IF
      borrow <> 0
        overflow()
      TRUE
        SKIP
:

PROC multiply([W]INT x, y)
  -- x := x*y
  [W]INT z, temp:
  INT carry, n, m:
  SEQ
    n := length(x)
    m := length(y)
    IF
      (n + m) <= W
        SEQ
          z := zero
          SEQ i = 0 FOR m
            SEQ
              carry := 0
              SEQ j = 0 FOR n
                carry, temp[j] := LONGPROD(x[j], y[i], carry)
              temp[n] := carry
              carry := 0
              SEQ j = i FOR n+1
                carry, z[j] := LONGSUM(temp[j-i], z[j], carry)
```

```
          IF
            z[w] = 0
              x := z
            TRUE
              overflow()
      TRUE
        overflow()
:

INT FUNCTION trial(VAL [W]INT r, d, VAL INT k, m)
  -- trial = min(r[k+m..k+m-1]/d[m-1], b - 1)
  VAL km IS k + m:
  INT qm, rm, test.hi, test.lo:
  VALOF
    SEQ
      qm, rm := LONGDIV(r[km], r[km-1], d[m-1])
      test.hi, test.lo := LONGPROD(qm, d[m-1], rm)
      IF
        (test.hi<>r[km]) OR (test.lo<>r[km-1])
          qm := b1
        TRUE
          SKIP
    RESULT qm
:

BOOL FUNCTION smaller(VAL [W]INT r, dq, VAL INT k, m)
  -- smaller = r[k+m..k] < dq[m..0]
  INT i, j, borrow, temp:
  VALOF
    SEQ
      i, j := m, 0
      WHILE i <> j
        IF
          r[i + k] <> dq[i]
            j := i
          TRUE
            i := i - 1
      borrow, temp := LONGDIFF(r[i+k], dq[i], 0)
    RESULT (borrow = 1)
:

PROC difference([W]INT r, dq, VAL INT k, m)
  -- r[k+m..k] := r[k+m..k] - dq[m..0]
  INT borrow, diff:
  SEQ
    borrow := 0
    SEQ i = 0 FOR m + 1
      borrow, r[i+k] := LONGDIFF(r[i+k], dq[i], borrow)
    IF
      borrow <> 0
        overflow()
      TRUE
        SKIP
:

PROC longdivide([W]INT x, y, q, r, VAL INT n, m)
  -- 2 <= m <= n <= w:
  -- q := x div y; r := x mod y
  INT f, k, qt, dummy:
  [W]INT d, dq, rf:
  SEQ
    IF
      y[m-1] < 0
        f := 1
```

```
        TRUE
          f, dummy := LONGDIV(1, 0, y[m-1]+1)
      product(r, x, f)
      product(d, y, f)
      q := zero
      k := n - m
      WHILE k >= 0
        SEQ
          qt := trial(r, d, k, m)
          product(dq, d, qt)
          WHILE smaller(r, dq, k, m)
            SEQ
              qt := qt - 1
              product(dq, d, qt)
          q[k] := qt
          difference(r, dq, k, m)
          k := k - 1
      rf := r
      quotient(r, rf, f)
:

PROC division([W]INT x, y, q, r)
  -- q := x div y; r := x mod y
  INT m, n, y1:
  SEQ
    m := length(y)
    IF
      m = 1
        SEQ
          y1 := y[m - 1]
          IF
            y1 <> 0
              SEQ
                quotient(q, x, y1)
                remainder(r, x, y1)
            TRUE
              overflow()
      TRUE
        SEQ
          n := length(x)
          IF
            m > n
              SEQ
                q := zero
                r := x
            TRUE
              longdivide(x, y, q, r, n, m)
:

PROC divide([W]INT x, y)
  -- x := x div y
  [W]INT q, r:
  SEQ
    division(x, y, q, r)
    x := q
:

PROC modulo([W]INT x, y)
  -- x := x mod y
  [W]INT q, r:
  SEQ
    division(x, y, q, r)
    x := r
:
```

```
PROC square([W]INT x)
  -- x := x*x
  [W]INT x0:
  IF
    (2*length(x)) <= w
      SEQ
        x0 := x
        multiply(x, x0)
    TRUE
      overflow()
:

PROC power([W]INT x, y)
  -- x := x**y
  [W]INT u, v:
  SEQ
    u := one
    v := y
    WHILE greater(v, zero)
      IF
        odd(v)
          SEQ
            multiply(u, x)
            decrement(v)
        TRUE
          SEQ
            square(x)
            halve(v)
    x := u
:

PROC modpower([W]INT x, y, z)
  -- x := (x**y) mod z
  [W]INT u, v:
  SEQ
    u := one
    v := y
    WHILE greater(v, zero)
      IF
        odd(v)
          SEQ
            multiply(u, x)
            modulo(u, z)
            decrement(v)
        TRUE
          SEQ
            square(x)
            modulo(x, z)
            halve(v)
    x := u
:

PROC randomdigit(INT x)
  -- 0 <= x <= b-1
  VAL a IS 16807(INT64):
  VAL m IS 2147483647(INT64):
  REAL64 f:
  SEQ
    seed := (a*seed) REM m
    f := (REAL64 ROUND seed)/(REAL64 ROUND m)
    x := INT TRUNC(f*(REAL64 ROUND b))
:
```

```
PROC randomlong([W]INT x, VAL INT digits)
  SEQ
    x := zero
    SEQ i = 0 FOR digits
      randomdigit(x[i])
:
PROC random([W]INT x, max)
  -- 1 <= x <= max
  SEQ
    randomlong(x, length(max))
    modulo(x, max)
    increment(x)
:

PROC witness([W]INT x, p, BOOL sure)
  [W]INT e, m, p1, r, y:
  SEQ
    m := one
    y := x
    e := p
    decrement(e)
    p1 := e
    sure := FALSE
    WHILE (NOT sure) AND greater(e, zero)
      IF
        odd(e)
          SEQ
            multiply(m, y)
            modulo(m, p)
            decrement(e)
        TRUE
          SEQ
            r := y
            square(y)
            modulo(y, p)
            halve(e)
            IF
              equal(y, one)
                sure := less(one, r) AND less(r, p1)
              TRUE
                SKIP
    sure := sure OR (NOT equal(m, ·one))
:

PROC randomdigit(INT x)
  -- 0 <= x <= b-1
  VAL a IS 16807(INT64):
  VAL m IS 2147483647(INT64):
  REAL64 f:
  SEQ
    seed := (a*seed) REM m
    f := (REAL64 ROUND seed)/(REAL64 ROUND m)
    x := INT TRUNC(f*(REAL64 ROUND b))
:

PROC randomlong([W]INT x, VAL INT digits)
  SEQ
    x := zero
    SEQ i = 0 FOR digits
      randomdigit(x[i])
:

PROC random([W]INT x, max)
  -- 1 <= x <= max
```

```
    SEQ
      randomlong(x, length(max))
      modulo(x, max)
      increment(x)
:

PROC initialize(VAL INT seed0)
  -- seed0 > 0
  SEQ
    seed := INT64(seed0)
    longint(zero, 0)
    longint(one, 1)
    longint(two, 2)
    SEQ i = 0 FOR w
      maxlong[i] := b1
    maxlong[w] := 0
:
```

Appendix 3

Strong Prime Generation

```
--        STRONG PRIME GENERATION
--            (Long Arithmetic)
--            4 November 1993
-- Copyright 1993, J. S. Greenfield
--
-- Certification based on Per Brinch Hansen's
-- primality testing program, copyright 1992.


-- The user procedure includes long arithmetic
-- and arithmetic tests defined in unlisted,
-- filed folds. The arithmetic fold is listed
-- in the element procedure.

VAL p IS 40:              -- processors
VAL m IS 40:              -- trials (m\p=0)
VAL logb IS 9:            -- approximate base ten log of radix
VAL lgb IS 32:            -- base 2 log of radix
VAL b IS 1000000000:      -- pseudo-radix (for random number generation)
VAL b1 IS -1:             -- highest digit (when converted to unsigned)
VAL N10 IS 200:           -- decimal test digits
VAL N IS 660:             -- binary test digits
VAL n IS (N/lgb)+1:       -- radix test digits
VAL w10 IS 2*N10:         -- max decimal digits
VAL w IS 2*n:             -- max radix digits
VAL W10 IS w10 + 1:       -- decimal array length
VAL W IS w + 1:           -- radix array length
VAL testing1 IS FALSE:    -- arithmetic test
VAL testing2 IS FALSE:    -- overflow test

PROTOCOL LONG
  CASE
    simple; [W]INT                  -- start searching for simple prime
    Double; [W]INT; INT             -- start searching for double prime
    strong; [W]INT; [W]INT; INT     -- start searching for strong prime
    result; [W]INT                  -- prime candidate
    check
    continue
    complete
    eos
    certify; [W]INT                 -- certify prime candidate
    Trial; BOOL                     -- witness
:
```

```
PROC USER(CHAN OF FILE inp0, out0, CHAN OF LONG inp1, out1, inp2, out2)
  [W]INT zero, one, two, maxlong:
  INT64 seed:

  PROC select([W]INT x1, x2)
    INT test, digits, seed0, prime.bits, prime.size:
    SEQ
      writestr("Test case [1, 2] = ")
      readint(test)
      IF
        test = 1
          INT s1, s2:
          SEQ
            writestr("s1 = ")
            readint(s1)
            longint(x1, s1)
            writestr("s2 = ")
            readint(s2)
            longint(x2, s2)
        test = 2
          [W10]INT y:
          SEQ
            writeln("Enter desired number of bits:")
            readint(prime.bits)
            prime.size := (prime.bits/lgb)/2
            writeln("Enter seed (0 < seed < 2**31-1):")
            readint(seed0)
            seed := INT64 seed0
            writeln("Random odd numbers:")
            randomlong(x1, prime.size)
            randomlong(x2, prime.size)
      IF
        NOT odd(x1)
          increment(x1)
        TRUE
          SKIP
      IF
        NOT odd(x2)
          increment(x2)
        TRUE
          SKIP
      writelong(x1)
      write(nl)
      write(nl)
      writelong(x2)
      write(nl)
  :

  PROC master([W]INT seed1, seed2, p, CHAN OF LONG inp1, out1, inp2, out2)

    PROC test(VAL [m]BOOL sure, BOOL prime)
      SEQ
        prime := TRUE
        SEQ i = 0 FOR m
          IF
            sure[i]
              prime := FALSE
            TRUE
              SKIP
    :

    PROC construct.p([W]INT p0, rs, r, s)
      [W]INT r1, s1, us, ur:
```

```
SEQ
  rs := r
  multiply(rs, s)
  r1 := r
  decrement(r1)
  s1 := s
  decrement(s1)
  us := s
  modpower(us, r1, rs)
  ur := r
  modpower(ur, s1, rs)
  add(us, rs)
  subtract(us, ur)
  modulo(us, rs)
  p0 := us
  IF
    NOT odd(us)
      add(p0, rs)
    TRUE
      SKIP
:

PROC GenerateSimplePrime ([W]INT seed, p,
  CHAN OF LONG inp1, out1, inp2, out2)
  BOOL prime:
  [m]BOOL b:
  SEQ
    prime := FALSE
    WHILE NOT prime
      SEQ
        out2!simple;seed
        inp2?CASE result;p
        out2!complete
        out1!certify;p
        SEQ i = 0 FOR m
          inp1?CASE Trial;b[i]
        test(b, prime)
        inp2?CASE complete
        seed := p
        sum(seed, 2)
:

PROC GenerateDoublePrime ([W]INT t, r,
  CHAN OF LONG inp1, out1, inp2, out2)
  [W]INT c:
  INT k:
  BOOL prime:
  [m]BOOL b:
  SEQ
    k := 0
    prime := FALSE
    WHILE NOT prime
      SEQ
        out2!Double;t;k
        inp2?CASE result;r
        out2!complete
        out1!certify;r
        SEQ i = 0 FOR m
          inp1?CASE Trial;b[i]
        test(b, prime)
        inp2?CASE complete
        c := r
        divide(c, t)
        k := shortint(c)
```

```
    :
    PROC GenerateStrongPrime ([W]INT r, s, p,
      CHAN OF LONG inp1, out1, inp2, out2)
      [W]INT rs, u, p0, c:
      INT k:
      BOOL prime:
      [m]BOOL b:
      SEQ
        k := 0
        construct.p(p0, rs, r, s)
        prime := FALSE
        WHILE NOT prime
          SEQ
            out2!strong;p0;rs;k
            inp2?CASE result;p
            out2!complete
            out1!certify;p
            SEQ i = 0 FOR m
              inp1?CASE Trial;b[i]
            test(b, prime)
            inp2?CASE complete
            c := p
            divide(c, rs)
            k := shortint(c)
    :

    [W]INT s, r, t:

    SEQ
      GenerateSimplePrime(seed1, s, inp1, out1, inp2, out2)
      GenerateSimplePrime(seed2, t, inp1, out1, inp2, out2)
      GenerateDoublePrime(t, r, inp1, out1, inp2, out2)
      GenerateStrongPrime(r, s, p, inp1, out1, inp2, out2)
      out2!eos
      out1!eos
    :

    PROC summarize([W]INT p)
      [W]INT temp:
      SEQ
        write(nl)
        write(nl)
        writeln("p =")
        writelong(p)
        write(nl)
        writeln("is a certified strong prime.")
        write(nl)
        write(nl)
    :

    PROC open(CHAN OF FILE inp, out)
      SEQ
        out!char;reallength
        out!int;12
        out!char;fraclength
        out!int;4
        out!char;intlength
        out!int;3
    :

    PROC close(VAL INT time, CHAN OF FILE inp, out)
      -- 1 second = 15625 units of 64 ms
      VAL units IS REAL64 ROUND time:
```

```
    VAL Tp IS units/15625.0(REAL64):
    VAL eps IS 0.00005(REAL64):
    SEQ
      out!char;nl
      out!str;18::" p      N      Tp (s)"
      out!char;nl
      out!char;nl
      out!int;p
      out!str;2::"   "
      out!int;N
      out!str;1::" "
      out!real;Tp + eps
      out!char;nl
      out!char;nl
      out!str;10::"type eof: "
      inp?CASE eof
      out!eof
  :

  [W]INT seed1, seed2, p:
  INT k:
  INT t1, t2:
  TIMER clock:
  SEQ
    open(inp0, out0)
    initialize(1)
    IF
      testing1
        longtest()
      TRUE
        SKIP
    IF
      testing2
        overflowtest()
      TRUE
        SKIP
    select(seed1, seed2)
    clock?t1
    master(seed1, seed2, p, inp1, out1, inp2, out2)
    clock?t2
    summarize(p)
    close(t2 MINUS t1, inp0, out0)
:

PROC ELEMENT(VAL INT id,
  CHAN OF LONG a.inp, a.out, c.inp, c.out, b.inp, b.out)

  VAL numprimes IS 95:
  VAL primes IS
    [2,3,5,7,11,13,17,19,23,29,31,37,41,43,47,53,59,61,67,71,73,79,83,89,97,
    101,103,107,109,113,127,131,137,139,149,151,157,163,167,173,179,181,191,
    193,197,199,211,223,227,229,233,239,241,251,257,263,269,271,277,281,283,
    293,307,311,313,317,331,337,347,349,353,359,367,373,379,383,389,397,401,
    409,419,421,431,443,433,439,449,457,461,463,467,479,487,491,499]:


  PROC generator(VAL INT id, CHAN OF LONG inp1, out1, inp2, out2)

    PROC witness([W]INT x, p, BOOL sure)
      [W]INT e, m, p1, r, y:
      SEQ
        m := one
        y := x
        e := p
```

```
        decrement(e)
        p1 := e
        sure := FALSE
        WHILE (NOT sure) AND greater(e, zero)
          IF
            odd(e)
              SEQ
                multiply(m, y)
                modulo(m, p)
                decrement(e)
            TRUE
              SEQ
                r := y
                square(y)
                modulo(y, p)
                halve(e)
                IF
                  equal(y, one)
                    sure := less(one, r) AND less(r, p1)
                  TRUE
                    SKIP
        sure := sure OR (NOT equal(m, one))
  :

  PROC prime([W]INT p, VAL INT m, BOOL pass)
    [W]INT x, p1:
    BOOL sure:
    SEQ
      p1 := p
      decrement(p1)
      pass := TRUE
      SEQ i = 1 FOR m
        SEQ
          random(x, p1)
          witness(x, p, sure)
          pass := pass AND (NOT sure)
  :

  BOOL FUNCTION SmallFactor(VAL [W]INT a)
    INT i, j, len, p, carry, dummy:
    BOOL factor:
    VALOF
      SEQ
        j := 1
        factor := FALSE
        len := length(a)-1
        WHILE (NOT factor) AND (j < numprimes)
          SEQ
            p := primes[j]
            carry := 0
            i := len
            WHILE i >= 0
              SEQ
                dummy, carry := LONGDIV(carry, a[i], p)
                i := i-1
            factor := (carry = 0)
            j := j+1
      RESULT factor
  :

  PROC searcher(VAL INT id, CHAN OF LONG inp1, out1, out2)

    PROC FindSimple(VAL INT id, [W]INT x, CHAN OF LONG inp1, out1, out2)
      [W]INT y:
```

```
INT k:
BOOL more, found:
SEQ
  k := (id-1)*2
  more := TRUE
  WHILE more
    SEQ
      out1!check
      inp1?CASE
        complete
          more := FALSE
        continue
          SEQ
            y := x
            sum(y, k)
            IF
              NOT SmallFactor(y)
                prime(y, 1, found)
              TRUE
                found := FALSE
            IF
              found
                SEQ
                  out2!result;y
                  out1!complete
                  more := FALSE
              TRUE
                SKIP
            k := k+(p*2)
  out2!complete
:

PROC FindDouble(VAL INT id, [W]INT x, INT k,
  CHAN OF LONG inp1, out1, out2)
  [W]INT y:
  BOOL more, found:
  SEQ
    k := k+(id*2)
    more := TRUE
    WHILE more
      SEQ
        out1!check
        inp1?CASE
          complete
            more := FALSE
          continue
            SEQ
              product(y, x, k)
              increment(y)
              IF
                NOT SmallFactor(y)
                  prime(y, 1, found)
                TRUE
                  found := FALSE
              IF
                found
                  SEQ
                    out2!result;y
                    out1!complete
                    more := FALSE
                TRUE
                  SKIP
              k := k+(p*2)
    out2!complete
```

```
    :

PROC FindStrong(VAL INT id, [W]INT p0, rs, INT k,
  CHAN OF LONG inp1, out1, out2)
  [W]INT y:
  BOOL more, found:
  SEQ
    k := k+(id*2)
    more := TRUE
    WHILE more
      SEQ
        out1!check
        inp1?CASE
          complete
            more := FALSE
          continue
            SEQ
              product(y, rs, k)
              add(y, p0)
              IF
                NOT SmallFactor(y)
                  prime(y, 1, found)
                TRUE
                  found := FALSE
              IF
                found
                  SEQ
                    out2!result;y
                    out1!complete
                    more := FALSE
                TRUE
                  SKIP
              k := k+(p*2)
    out2!complete
    :

[W]INT x, y:
INT k:
BOOL more:
SEQ
  more := TRUE
  WHILE more
    inp1?CASE
      simple;x
        FindSimple(id, x, inp1, out1, out2)
      Double;x;k
        FindDouble(id, x, k, inp1, out1, out2)
      strong;x;y;k
        FindStrong(id, x, y, k, inp1, out1, out2)
      eos
        SEQ
          out2!eos
          more := FALSE
    :

PROC reader(VAL INT id, CHAN OF LONG inp1, out1, inp2, out2)

  PROC monitor(VAL INT id, CHAN OF LONG inp1, out1, inp2, out2)
    BOOL more:
    SEQ
      more := TRUE
      WHILE more
        ALT
          inp1?CASE complete
```

```
              SEQ
                IF
                  id < p
                    out1!complete
                  TRUE
                    SKIP
                inp2?CASE
                  check
                    out2!complete
                  complete
                    SKIP
                more := FALSE
          inp2?CASE
            check
              out2!continue
            complete
              SEQ
                IF
                  id < p
                    out1!complete
                  TRUE
                    SKIP
                inp1?CASE complete
                more := FALSE
:

[W]INT x, y:
INT k:
BOOL more:
SEQ
  more := TRUE
  WHILE more
    inp1?CASE
      simple;x
        SEQ
          IF
            id < p
              out1!simple;x
            TRUE
              SKIP
          out2!simple;x
          monitor(id, inp1, out1, inp2, out2)
      Double;x;k
        SEQ
          IF
            id < p
              out1!Double;x;k
            TRUE
              SKIP
          out2!Double;x;k
          monitor(id, inp1, out1, inp2, out2)
      strong;x;y;k
        SEQ
          IF
            id < p
              out1!strong;x;y;k
            TRUE
              SKIP
          out2!strong;x;y;k
          monitor(id, inp1, out1, inp2, out2)
      eos
        SEQ
          IF
            id < p
```

```
                    out1!eos
                  TRUE
                    SKIP
                out2!eos
                more := FALSE
:

PROC writer(VAL INT id, CHAN OF LONG inp1, inp2, out2)

  PROC PassResult(VAL INT terminate, CHAN OF LONG inp1, inp2, out2,
    BOOL done)
    [W]INT y:
    INT count:
    BOOL passing:
    SEQ
      done := FALSE
      passing := TRUE
      count := 0
      WHILE (NOT done) AND (count < terminate)
        ALT
          inp1?CASE
            result;y
              IF
                passing
                  SEQ
                    out2!result;y
                    passing := FALSE
                TRUE
                  SKIP
            complete
              SEQ
                count := count+1
                IF
                  count = terminate
                    out2!complete
                  TRUE
                    SKIP
            eos
              done := TRUE
          inp2?CASE
            result;y
              IF
                passing
                  SEQ
                    out2!result;y
                    passing := FALSE
                TRUE
                  SKIP
            complete
              SEQ
                count := count+1
                IF
                  count = terminate
                    out2!complete
                  TRUE
                    SKIP
  :

  INT terminate:
  BOOL done:
  SEQ
    IF
      id < p
        terminate := 2
```

```
          TRUE
            terminate := 1
        done := FALSE
        WHILE NOT done
          PassResult(terminate, inp1, inp2, out2, done)
    :

  CHAN OF LONG a, b, c:

  SEQ
    initialize((id*id)*id)
    PAR
      reader(id, inp1, out1, a, b)
      searcher(id, b, a, c)
      writer(id, c, inp2, out2)
:

PROC certifier(VAL INT id, CHAN OF LONG inp, out)
  -- 1 <= id <= p
  VAL q IS m/p: -- quota (trials/processor) for certification
  [W]INT zero, one, two, maxlong:
  INT64 seed:

  -- Long arithmetic and primality tests are contained inside
  -- hidden folds.


  PROC solve([W]INT p, BOOL sure)
    [W]INT p1, x:
    SEQ
      p1 := p
      decrement(p1)
      random(x, p1)
      witness(x, p, sure)
  :

  [W]INT x:
  BOOL more, b:
  SEQ
    initialize(id*id)
    more := TRUE
    WHILE more
      inp?CASE
        certify;x
          SEQ
            IF
              id < p
                out!certify;x
              TRUE
                SKIP
            SEQ j = 1 FOR q
              SEQ
                solve(x, b)
                out!Trial;b
                SEQ k = 1 FOR id - 1
                  SEQ
                    inp?CASE Trial;b
                    out!Trial;b
        eos
          SEQ
            IF
              id < p
                out!eos
              TRUE
```

```
                    SKIP
                more := FALSE
      :

    PRI PAR
      certifier(id, a.inp, a.out)
      generator(id, c.inp, c.out, b.inp, b.out)
  :

PROC FILES(CHAN OF ANY inp1, out1, CHAN OF FILE inp2, out2)
  #USE "/iolib/io.lib"
  IO(out1, inp1, out2, inp2)
  :

CHAN OF ANY anyinp, anyout:
CHAN OF FILE fileinp, fileout:
[p + 1]CHAN OF LONG a, b, c:
PLACED PAR
  PROCESSOR 0 T8
    PLACE fileout AT inp0:
    PLACE fileinp AT out0:
    PLACE anyinp AT inp2:
    PLACE anyout AT out2:
    FILES(anyinp, anyout, fileout, fileinp)
  PROCESSOR 1 T8
    PLACE fileinp AT inp0:
    PLACE fileout AT out0:
    PLACE a[p] AT inp1:
    PLACE a[0] AT out2:
    PLACE b[0] AT inp3:
    PLACE c[0] AT out3:
    USER(fileinp, fileout, a[p], a[0], b[0], c[0])
  PLACED PAR i = 1 FOR p
    PROCESSOR i + 1 T8
      PLACE a[i - 1] AT inp0:
      PLACE a[i] AT out1:
      PLACE c[i-1] AT inp2:
      PLACE b[i-1] AT out2:
      PLACE b[i] AT inp3:
      PLACE c[i] AT out3:
      ELEMENT(i, a[i - 1], a[i],
        c[i-1], c[i], b[i], b[i-1])
```

Appendix 4

Pollard $p-1$ Factoring

```
--       Pollard p-1 Factoring
--          (Long Arithmetic)
--           28 October 1993
-- Copyright 1993 J. S. Greenfield

-- The user procedure includes long arithmetic
-- and arithmetic tests defined in unlisted,
-- filed folds. The arithmetic fold is listed
-- in the element procedure.

VAL p IS 40:               -- processors
VAL testSeconds IS 6000:   -- number of seconds before timeout
VAL limit IS 10000:
VAL logb IS 9:             -- approximate base ten log of radix
VAL lgb IS 32:            .-- base 2 log of radix
VAL b IS 1000000000:       -- pseudo-radix (for random number generation)
VAL b1 IS -1:              -- highest digit (when converted to unsigned)
VAL N10 IS 200:            -- decimal test digits
VAL N IS 660:              -- binary test digits
VAL n IS (N/lgb)+1:        -- radix test digits
VAL w10 IS 2*N10:          -- max decimal digits
VAL w IS 2*n:              -- max radix digits
VAL W10 IS w10 + 1:        -- decimal array length
VAL W IS w + 1:            -- radix array length
VAL testing1 IS FALSE:     -- arithmetic test
VAL testing2 IS FALSE:     -- overflow test

PROTOCOL LONG
  CASE
    start; [W]INT         -- composite to be factored
    result; [W]INT        -- factor
    check
    continue
    complete
    eos
:

PROC USER(CHAN OF FILE inp0, out0, CHAN OF LONG inp1, out1)
  [W]INT zero, one, two, maxlong:

  PROC ReadProduct([W]INT x)
    SEQ i=0 FOR W
      inp0?CASE int;x[i]
```

```
  :

PROC summarize([W]INT n, g, VAL BOOL found)
  SEQ
    write(nl)
    write(nl)
    writeln("n =")
    writelong(n)
    write(nl)
    write(nl)
    IF
      found
        SEQ
          writeln("A factor, g, was found.")
          writeln("g =")
          writelong(g)
          write(nl)
      TRUE
        writeln("No factor was found.")
    write(nl)
    write(nl)
    write(nl)
  :

PROC master([W]INT n, VAL INT timeLimit, [W]INT g, BOOL found,
  CHAN OF LONG inp, out)
  TIMER clock:
  INT startTime:
  SEQ
    out!start;n
    clock?startTime
    writeln("searching for factors...")
    found := TRUE
    ALT
      inp?CASE result;g
        SEQ
          out!complete
          inp?CASE complete
      clock?AFTER (startTime PLUS timeLimit)
        SEQ
          out!complete
          inp?CASE
            result;g
              inp?CASE complete
            complete
              found := FALSE
    out!eos
  :

PROC open(CHAN OF FILE inp, out)
  SEQ
    out!char;reallength
    out!int;12
    out!char;fraclength
    out!int;4
    out!char;intlength
    out!int;3
  :

PROC close(VAL INT time, CHAN OF FILE inp, out)
  -- 1 second = 15625 units of 64 ms
  VAL units IS REAL64 ROUND time:
  VAL Tp IS units/15625.0(REAL64):
  VAL eps IS 0.00005(REAL64):
```

```
    SEQ
      out!char;nl
      out!str;18::"  p     N     Tp (s)"
      out!char;nl
      out!char;nl
      out!int;p
      out!str;2::"  "
      out!int;N
      out!str;1::" "
      out!real;Tp + eps
      out!char;nl
      out!char;nl
      out!str;10::"type eof: "
      inp?CASE eof
      out!eof
  :

  VAL timeLimit IS testSeconds*15625:
  [W]INT n, g:
  BOOL found:
  INT t1, t2:
  TIMER clock:
  SEQ
    open(inp0, out0)
    initialize()
    IF
      testing1
        longtest()
      TRUE
        SKIP
    IF
      testing2
        overflowtest()
      TRUE
        SKIP
    ReadProduct(n)
    clock?t1
    master(n, timeLimit, g, found, inp1, out1)
    clock?t2
    summarize(n, g, found)
    close(t2 MINUS t1, inp0, out0)
:

PROC server(VAL INT id, CHAN OF LONG inp1, out1, inp2, out2)

  PROC searcher(VAL INT id, CHAN OF LONG inp1, out1, out2)

    PROC FindFactor(VAL INT id, [W]INT n, CHAN OF LONG inp1, out1, out2)
      BOOL more, found:
      INT c:
      [W]INT g, m, k:

      PROC initialize(INT c, [W]INT m, k, g, VAL INT id)
        SEQ
          c := id+1
          longint(m, c)
          k := one
          g := one
      :

      PROC factor([W]INT n, m, k, g, BOOL found)
        [W]INT mless1:
        SEQ
          modpower(m, k, n)
```

```
      IF
        greater(m, zero)
          SEQ
            mless1 := m
            decrement(mless1)
            gcd(g, mless1, n)
        TRUE
          g := zero
      increment(k)
      found := greater(g, one) AND less(g, n)
  :

  PROC update([W]INT n, INT c, [W]INT m, k, g)
    IF
      (shortint(k) > limit) OR (NOT equal(g, one))
        SEQ
          c := c+p
          longint(m, c)
          k := one
          g := one
      TRUE
        SKIP
  :

  SEQ
    initialize(c, m, k, g, id)
    more := TRUE
    WHILE more
      SEQ
        out1!check
        inp1?CASE
          complete
            more := FALSE
          continue
            SEQ
              factor(n, m, k, g, found)
              IF
                found
                  SEQ
                    out2!result;g
                    out1!complete
                    more := FALSE
                TRUE
                  SKIP
              update(n, c, m, k, g)
    out2!complete
  :

  [W]INT x:
  BOOL more:
  SEQ
    more := TRUE
    WHILE more
      inp1?CASE
        start;x
          FindFactor(id, x, inp1, out1, out2)
        eos
          SEQ
            out2!eos
            more := FALSE
  :

PROC reader(VAL INT id, CHAN OF LONG inp1, out1, inp2, out2)
```

```
PROC monitor(VAL INT id, CHAN OF LONG inp1, out1, inp2, out2)
  BOOL more:
  SEQ
    more := TRUE
    WHILE more
      ALT
        inp1?CASE complete
          SEQ
            IF
              id < p
                out1!complete
              TRUE
                SKIP
            inp2?CASE
              check
                out2!complete
              complete
                SKIP
            more := FALSE
        inp2?CASE
          check
            out2!continue
          complete
            SEQ
              IF
                id < p
                  out1!complete
                TRUE
                  SKIP
              inp1?CASE complete
              more := FALSE
  :

[W]INT x:
BOOL more:
SEQ
  more := TRUE
  WHILE more
    inp1?CASE
      start;x
        SEQ
          IF
            id < p
              out1!start;x
            TRUE
              SKIP
          out2!start;x
          monitor(id, inp1, out1, inp2, out2)
      eos
        SEQ
          IF
            id < p
              out1!eos
            TRUE
              SKIP
          out2!eos
          more := FALSE
:

PROC writer(VAL INT id, CHAN OF LONG inp1, inp2, out2)

  PROC PassResult(VAL INT terminate, CHAN OF LONG inp1, inp2, out2,
    BOOL done)
    [W]INT y:
```

```
INT count:
BOOL passing:
SEQ
  done := FALSE
  passing := TRUE
  count := 0
  WHILE (NOT done) AND (count < terminate)
    ALT
      inp1?CASE
        result;y
          IF
            passing
              SEQ
                out2!result;y
                passing := FALSE
            TRUE
              SKIP
        complete
          SEQ
            count := count+1
            IF
              count = terminate
                out2!complete
              TRUE
                SKIP
        eos
          done := TRUE
      inp2?CASE
        result;y
          IF
            passing
              SEQ
                out2!result;y
                passing := FALSE
            TRUE
              SKIP
        complete
          SEQ
            count := count+1
            IF
              count = terminate
                out2!complete
              TRUE
                SKIP
:

INT terminate:
BOOL done:
SEQ
  IF
    id < p
      terminate := 2
    TRUE
      terminate := 1
  done := FALSE
  WHILE NOT done
    PassResult(terminate, inp1, inp2, out2, done)
:

CHAN OF LONG a, b, c:

SEQ
  initialize()
  PAR
```

```
      reader(id, inp1, out1, a, b)
      searcher(id, b, a, c)
      writer(id, c, inp2, out2)
:

PROC FILES(CHAN OF ANY inp1, out1, CHAN OF FILE inp2, out2)
  #USE "/iolib/io.lib"
  IO(out1, inp1, out2, inp2)
:

CHAN OF ANY anyinp, anyout:
CHAN OF FILE fileinp, fileout:
[p + 1]CHAN OF LONG a, b:
PLACED PAR
  PROCESSOR 0 T8
    PLACE fileout AT inp0:
    PLACE fileinp AT out0:
    PLACE anyinp AT inp2:
    PLACE anyout AT out2:
    FILES(anyinp, anyout, fileout, fileinp)
  PROCESSOR 1 T8
    PLACE fileinp AT inp0:
    PLACE fileout AT out0:
    PLACE a[0] AT inp1:
    PLACE b[0] AT out1:
    USER(fileinp, fileout, a[0], b[0])
  PLACED PAR i = 1 FOR p
    PROCESSOR i + 1 T8
      PLACE b[i-1] AT inp1:
      PLACE a[i-1] AT out1:
      PLACE a[i] AT inp2:
      PLACE b[i] AT out2:
      server(i, b[i-1], b[i], a[i], a[i-1])
```

Appendix 5

RSA Enciphering

```
--         RSA   ENCIPHERING
--         24 November 1993
-- Copyright (c) 1993 J. S. Greenfield


VAL p IS 40:              -- processors
VAL radix.bits IS 32:     -- base 2 log of radix (number of bits in radix)
VAL b1 IS -1:             -- maximum digit (when converted to unsigned)
VAL N IS 20:              -- base-radix digits in modulus and exponent
VAL w IS 2*N:             -- max radix digits
VAL W IS w + 1:           -- radix array length

PROTOCOL LONG
  CASE
    global; [N]INT; [N]INT  -- an RSA exponent and modulus, <e, M>
    data; [N]INT            -- a chained plaintext block
    result; [N]INT          -- a ciphertext block
    complete
    eos
:

PROC USER(CHAN OF FILE inp0, out0, CHAN OF LONG inp1, out1)
  [N]INT zero:

  PROC ReadKey([N]INT e, M)
    SEQ
      SEQ i = 0 FOR N
        inp0?CASE int;M[i]
      SEQ i = 0 FOR N
        inp0?CASE int;e[i]
  :

  PROC ReadBlock([N]INT x, BOOL received, more)
    INT i:
    SEQ
      x := zero
      i := N-1
      more := TRUE
      WHILE more AND (i >= 0)
        SEQ
          inp0?CASE
            int;x[i]
              i := i-1
```

```
            eof
              more := FALSE
      received := (i < (N-1))
  :

PROC WriteBlock([N]INT x)
  SEQ i = 1 FOR N
    SEQ
      out0!int;x[N-i]
      out0!char;sp
  :

PROC distributor(CHAN OF LONG out)

  PROC chain([N]INT ui, ulast, mi)
    SEQ
      SEQ j = 0 FOR N
        ui[j] := ulast[j] >< mi[j]
      ulast := ui
    :

  BOOL more, received:
  INT i:
  [p][N]INT mi:
  [N]INT e, M, ulast, ui:
  SEQ
    ReadKey(e, M)
    out!global;e;M
    ulast := zero
    more := TRUE
    WHILE more
      SEQ
        i := 0
        WHILE more AND (i < p)
          SEQ
            ReadBlock(mi[i], received, more)
            IF
              received
                i := i+1
              TRUE
                SKIP
        SEQ j = 0 FOR i
          SEQ
            chain(ui, ulast, mi[j])
            out!data;ui
    out!complete
    out!eos
  :

PROC collector(CHAN OF LONG inp)
  BOOL more:
  INT i:
  [p][N]INT ci:
  SEQ
    more := TRUE
    WHILE more
      SEQ
        i := 0
        WHILE more AND (i < p)
          inp?CASE
            result;ci[i]
              i := i+1
            complete
              more := FALSE
```

```
          SEQ j = 0 FOR i
            WriteBlock(ci[j])
      inp?CASE eos
  :

  PROC initialize()
    SEQ i = 0 FOR N
      zero[i] := 0
  :

  PROC open(CHAN OF FILE inp, out)
    SEQ
      out!char;reallength
      out!int;12
      out!char;fraclength
      out!int;4
      out!char;intlength
      out!int;3
  :

  PROC close(VAL INT time, CHAN OF FILE inp, out)
    -- 1 second = 15625 units of 64 ms
    VAL units IS REAL64 ROUND time:
    VAL Tp IS units/15625.0(REAL64):
    SEQ
      out!char;nl
      out!char;nl
      out!char;nl
      out!str;16::" p      N    Tp (s)"
      out!char;nl
      out!char;nl
      out!int;p
      out!str;2::"  "
      out!int;N
      out!str;1::" "
      out!real;Tp
      out!char;nl
      out!char;nl
      out!eof
  :

  INT t1, t2:
  TIMER clock:
  SEQ
    initialize()
    open(inp0, out0)
    clock?t1
    PAR
      distributor(out1)
      collector(inp1)
    clock?t2
    close(t2 MINUS t1, inp0, out0)
:

PROC ELEMENT(VAL INT id, CHAN OF LONG inp, out)

  PROC server(VAL INT id, CHAN OF LONG inp, out)

    PROC broadcast([W]INT e, M, BOOL more, VAL INT id,
      CHAN OF LONG inp, out)
      [N]INT e1, M1:
      inp?CASE
        global;e1;M1
          SEQ
```

```
        IF
          id > 1
            out!global;e1;M1
          TRUE
            SKIP
        e := zero
        SEQ i = 0 FOR N
          e[i] := e1[i]
        M := zero
        SEQ i = 0 FOR N
          M[i] := M1[i]
        more := TRUE
    eos
      SEQ
        out!eos
        more := FALSE
:

PROC ShiftIn([W]INT ui, BOOL more, VAL INT id, CHAN OF LONG inp, out)
  [N]INT a:
  INT i:
  SEQ
    i := 1
    more := TRUE
    WHILE more AND (i < id)
      inp?CASE
        data;a
          SEQ
            out!data;a
            i := i+1
        complete
          SEQ
            out!complete
            more := FALSE
    IF
      more
        inp?CASE
          data;a
            SEQ
              ui := zero
              SEQ j = 0 FOR N
                ui[j] := a[j]
          complete
            SEQ
              out!complete
              more := FALSE
      TRUE
        SKIP
:

PROC ShiftOut([W]INT ci, BOOL more, VAL INT id, CHAN OF LONG inp, out)
  [N]INT b:
  INT i:
  SEQ
    SEQ j = 0 FOR N
      b[j] := ci[j]
    out!result;b
    i := id+1
    more := TRUE
    WHILE more AND (i <= p)
      SEQ
        inp?CASE
          result;b
            SEQ
```

```
                          out!result;b
                          i := i+1
                    complete
                      SEQ
                        out!complete
                        more := FALSE
        :

    [W]INT e, M, ui, ci:
    BOOL continue, more:
    SEQ
      initialize()
      broadcast(e, M, continue, id, inp, out)
      WHILE continue
        SEQ
          more := TRUE
          WHILE more
            SEQ
              ShiftIn(ui, more, id, inp, out)
              IF
                more
                  SEQ
                    ci := ui
                    modpower(ci, e, M)
                    ShiftOut(ci, more, id, inp, out)
                TRUE
                  SKIP
          broadcast(e, M, continue, id, inp, out)
    :

  PROC copy(CHAN OF LONG inp, out)
    BOOL more:
    [N]INT x, y:
    SEQ
      more := TRUE
      WHILE more
        inp?CASE
          global;x;y
            out!global;x;y
          data;x
            out!data;x
          result;x
            out!result;x
          complete
            out!complete
          eos
            SEQ
              out!eos
              more := FALSE
    :

  CHAN OF LONG a, b:
  PAR
    copy(inp, a)
    copy(b, out)
    server(id, a, b)
  :

PROC FILES(CHAN OF ANY inp1, out1, CHAN OF FILE inp2, out2)
  #USE "/iolib/io.lib"
  IO(out1, inp1, out2, inp2)
:

CHAN OF ANY anyinp, anyout:
```

```
CHAN OF FILE fileinp, fileout:
[p + 1]CHAN OF LONG c:
PLACED PAR
  PROCESSOR 0 T8
    PLACE fileout AT inp0:
    PLACE fileinp AT out0:
    PLACE anyinp AT inp2:
    PLACE anyout AT out2:
    FILES(anyinp, anyout, fileout, fileinp)
  PROCESSOR 1 T8
    PLACE fileinp AT inp0:
    PLACE fileout AT out0:
    PLACE c[0] AT inp1:
    PLACE c[p] AT out2:
    USER(fileinp, fileout, c[0], c[p])
  PLACED PAR i = 1 FOR p
    PROCESSOR i + 1 T8
      PLACE c[i] AT inp0:
      PLACE c[i-1] AT out1:
      ELEMENT(i, c[i], c[i-1])
```

Appendix 6

RSA Deciphering

```
--        RSA   DECIPHERING
--        24 November 1993
-- Copyright (c) 1993 J. S. Greenfield


VAL p IS 40:              -- processors
VAL radix.bits IS 32:     -- base 2 log of radix (number of bits in radix)
VAL b1 IS -1:             -- maximum digit (when converted to unsigned)
VAL N IS 20:              -- base-radix digits in modulus and exponent
VAL w IS 2*N:             -- max radix digits
VAL W IS w + 1:           -- radix array length

PROTOCOL LONG
  CASE
    global; [N]INT; [N]INT  -- an RSA exponent and modulus, <e, M>
    data; [N]INT            -- a chained plaintext block
    result; [N]INT          -- a ciphertext block
    complete
    eos
:

PROC USER(CHAN OF FILE inp0, out0, CHAN OF LONG inp1, out1)
  [N]INT zero:

  PROC ReadKey([N]INT d, M)
    SEQ
      SEQ i = 0 FOR N
        inp0?CASE int;M[i]
      SEQ i = 0 FOR N
        inp0?CASE int;d[i]
  :

  PROC ReadBlock([N]INT x, BOOL received, more)
    INT i:
    SEQ
      x := zero
      i := N-1
      more := TRUE
      WHILE more AND (i >= 0)
        SEQ
          inp0?CASE
            int;x[i]
              i := i-1
```

```
                eof
                  more := FALSE
         received := (i < (N-1))
  :

PROC WriteBlock([N]INT x)
  SEQ i = 1 FOR N
    SEQ
      out0!int;x[N-i]
      out0!char;sp
  :

PROC distributor(CHAN OF LONG out)
  BOOL more, received:
  INT i:
  [p][N]INT ci:
  [N]INT d, M:
  SEQ
    ReadKey(d, M)
    out!global;d;M
    more := TRUE
    WHILE more
      SEQ
        i := 0
        WHILE more AND (i < p)
          SEQ
            ReadBlock(ci[i], received, more)
            IF
              received
                i := i+1
              TRUE
                SKIP
        SEQ j = 0 FOR i
          out!data;ci[j]
    out!complete
    out!eos
  :

PROC collector(CHAN OF LONG inp)

  PROC unchain([N]INT mi, ulast, ui)
    SEQ
      SEQ j = 0 FOR N
        mi[j] := ulast[j] >< ui[j]
      ulast := ui
    :

  BOOL more:
  INT i:
  [p][N]INT ui:
  [N]INT ulast, mi:
  SEQ
    ulast := zero
    more := TRUE
    WHILE more
      SEQ
        i := 0
        WHILE more AND (i < p)
          inp?CASE
            result;ui[i]
              i := i+1
            complete
              more := FALSE
        SEQ j = 0 FOR i
```

```
            SEQ
              unchain(mi, ulast, ui[j])
              WriteBlock(mi)
      inp?CASE eos
  :

  PROC initialize()
    SEQ i = 0 FOR N
      zero[i] := 0
  :

  PROC open(CHAN OF FILE inp, out)
    SEQ
      out!char;reallength
      out!int;12
      out!char;fraclength
      out!int;4
      out!char;intlength
      out!int;3
  :

  PROC close(VAL INT time, CHAN OF FILE inp, out)
    -- 1 second = 15625 units of 64 ms
    VAL units IS REAL64 ROUND time:
    VAL Tp IS units/15625.0(REAL64):
    SEQ
      out!char;nl
      out!char;nl
      out!char;nl
      out!str;16::" p     N    Tp (s)"
      out!char;nl
      out!char;nl
      out!int;p
      out!str;2::"  "
      out!int;N
      out!str;1::" "
      out!real;Tp
      out!char;nl
      out!char;nl
      out!eof
  :

  INT t1, t2:
  TIMER clock:
  SEQ
    initialize()
    open(inp0, out0)
    clock?t1
    PAR
      distributor(out1)
      collector(inp1)
    clock?t2
    close(t2 MINUS t1, inp0, out0)
:

PROC ELEMENT(VAL INT id, CHAN OF LONG inp, out)

  PROC server(VAL INT id, CHAN OF LONG inp, out)

    PROC broadcast([W]INT d, M, BOOL more, VAL INT id,
      CHAN OF LONG inp, out)
      [N]INT d1, M1:
      inp?CASE
        global;d1;M1
```

```
      SEQ
        IF
          id > 1
            out!global;d1;M1
          TRUE
            SKIP
        d := zero
        SEQ i = 0 FOR N
          d[i] := d1[i]
        M := zero
        SEQ i = 0 FOR N
          M[i] := M1[i]
        more := TRUE
    eos
      SEQ
        out!eos
        more := FALSE
:

PROC ShiftIn([W]INT ci, BOOL more, VAL INT id, CHAN OF LONG inp, out)
  [N]INT a:
  INT i:
  SEQ
    i := 1
    more := TRUE
    WHILE more AND (i < id)
      inp?CASE
        data;a
          SEQ
            out!data;a
            i := i+1
        complete
          SEQ
            out!complete
            more := FALSE
    IF
      more
        inp?CASE
          data;a
            SEQ
              ci := zero
              SEQ j = 0 FOR N
                ci[j] := a[j]
          complete
            SEQ
              out!complete
              more := FALSE
      TRUE
        SKIP
:

PROC ShiftOut([W]INT ui, BOOL more, VAL INT id, CHAN OF LONG inp, out)
  [N]INT b:
  INT i:
  SEQ
    SEQ j = 0 FOR N
      b[j] := ui[j]
    out!result;b
    i := id+1
    more := TRUE
    WHILE more AND (i <= p)
      SEQ
        inp?CASE
          result;b
```

```
                    SEQ
                      out!result;b
                      i := i+1
                  complete
                    SEQ
                      out!complete
                      more := FALSE
      :

    [W]INT d, M, ci, ui:
    BOOL continue, more:
    SEQ
      initialize()
      broadcast(d, M, continue, id, inp, out)
      WHILE continue
        SEQ
          more := TRUE
          WHILE more
            SEQ
              ShiftIn(ci, more, id, inp, out)
              IF
                more
                  SEQ
                    ui := ci
                    modpower(ui, d, M)
                    ShiftOut(ui, more, id, inp, out)
                TRUE
                  SKIP
          broadcast(d, M, continue, id, inp, out)
  :

  PROC copy(CHAN OF LONG inp, out)
    BOOL more:
    [N]INT x, y:
    SEQ
      more := TRUE
      WHILE more
        inp?CASE
          global;x;y
            out!global;x;y
          data;x
            out!data;x
          result;x
            out!result;x
          complete
            out!complete
          eos
            SEQ
              out!eos
              more := FALSE
  :

  CHAN OF LONG a, b:
  PAR
    copy(inp, a)
    copy(b, out)
    server(id, a, b)
:

PROC FILES(CHAN OF ANY inp1, out1, CHAN OF FILE inp2, out2)
  #USE "/iolib/io.lib"
  IO(out1, inp1, out2, inp2)
:
```

```
CHAN OF ANY anyinp, anyout:
CHAN OF FILE fileinp, fileout:
[p + 1]CHAN OF LONG c:
PLACED PAR
  PROCESSOR 0 T8
    PLACE fileout AT inp0:
    PLACE fileinp AT out0:
    PLACE anyinp AT inp2:
    PLACE anyout AT out2:
    FILES(anyinp, anyout, fileout, fileinp)
  PROCESSOR 1 T8
    PLACE fileinp AT inp0:
    PLACE fileout AT out0:
    PLACE c[0] AT inp1:
    PLACE c[p] AT out2:
    USER(fileinp, fileout, c[0], c[p])
  PLACED PAR i = 1 FOR p
    PROCESSOR i + 1 T8
      PLACE c[i] AT inp0:
      PLACE c[i-1] AT out1:
      ELEMENT(i, c[i], c[i-1])
```

Appendix 7

Deterministic Prime Certification

```
-- LUCAS-CERTIFIED PRIME GENERATION
--        (Long Arithmetic)
--        24 November 1993
-- Copyright 1993, J. S. Greenfield


-- The user procedure includes long arithmetic
-- and arithmetic tests defined in unlisted,
-- filed folds. The arithmetic fold is listed
-- in the element procedure.

VAL p IS 40:              -- processors
VAL logb IS 9:            -- approximate base ten log of radix
VAL lgb IS 32:            -- base 2 log of radix
VAL b IS 1000000000:      -- pseudo-radix (for random number generation)
VAL b1 IS -1:             -- highest digit (when converted to unsigned)
VAL N10 IS 200:           -- decimal test digits
VAL N IS 640:             -- binary test digits
VAL n IS (N/lgb)+1:       -- radix test digits
VAL w10 IS 2*N10:         -- max decimal digits
VAL w IS 2*n:             -- max radix digits
VAL W10 IS w10 + 1:       -- decimal array length
VAL W IS w + 1:           -- radix array length
VAL testing1 IS FALSE:    -- arithmetic test
VAL testing2 IS FALSE:    -- overflow test (N=2,b=10)

VAL numprimes IS 95:
VAL primes IS
  [2,3,5,7,11,13,17,19,23,29,31,37,41,43,47,53,59,61,67,71,73,79,83,89,97,
   101,103,107,109,113,127,131,137,139,149,151,157,163,167,173,179,181,191,
   193,197,199,211,223,227,229,233,239,241,251,257,263,269,271,277,281,283,
   293,307,311,313,317,331,337,347,349,353,359,367,373,379,383,389,397,401,
   409,419,421,431,443,433,439,449,457,461,463,467,479,487,491,499]:


PROTOCOL LONG
  CASE
    start; INT; [W]INT                 -- start searching for a prime
    candidate; [W]INT; [numprimes]BOOL -- prime candidate with factors
    check
    continue
    complete
    eos
```

```
    global; [W]INT                      -- prime candidate to be certified
    data; [W]INT                        -- a factor for primality testing
    result;BOOL                         -- result of primality test
:

PROC USER(CHAN OF FILE inp0, out0, CHAN OF LONG a.inp, a.out, b.inp, c.out)

  VAL bitsLeft IS 32:

  PROC select(INT seed0, bits)
    SEQ
      writeln("Enter seed (0 < seed < 2**31-1):")
      readint(seed0)
      writeln("Enter desired number of bits:")
      readint(bits)
      write(nl)
  :

  PROC RandomProduct(VAL INT bits, [W]INT n, [numprimes]BOOL factor)
    [W]INT x:
    INT i:
    SEQ
      n := two
      factor[0] := TRUE
      SEQ i = 1 FOR numprimes-1
        factor[i] := FALSE
      WHILE size(n) < bits
        SEQ
          randomdigit(i)
          i := i REM numprimes
          product(x, n, primes[i])
          IF
            size(x) <= bits
              SEQ
                n := x
                factor[i] := TRUE
            TRUE
              SKIP
  :

  PROC master(VAL INT bits, [W]INT n, CHAN OF LONG inp1, out1, inp2, out2)

    PROC certify([W]INT n, [numprimes]BOOL factor, BOOL pass,
      CHAN OF LONG inp, out)
      [W]INT q:
      SEQ
        out!global;n
        SEQ i = 0 FOR numprimes
          IF
            factor[i]
              SEQ
                longint(q, primes[i])
                out!data;q
            TRUE
              SKIP
        out!complete
        inp?CASE result;pass
    :

    [W]INT x:
    [numprimes]BOOL factor, f:
    BOOL prime:
    SEQ
      RandomProduct(bits-bitsLeft, x, f)
```

```
        prime := FALSE
        WHILE NOT prime
          SEQ
            out2!start;bits;x
            inp2?CASE candidate;n;factor
            SEQ i = 0 FOR numprimes
              factor[i] := factor[i] AND f[i]
            out2!complete
            certify(n, factor, prime, inp1, out1)
            inp2?CASE complete
        out2!eos
        out1!eos
  :

  PROC collector(CHAN OF LONG inp, out)
    BOOL more, prime, pass:
    SEQ
      prime := TRUE
      more := TRUE
      WHILE more
        SEQ
          inp?CASE
            result;pass
              prime := prime AND pass
            complete
              SEQ
                out!result;prime
                prime := TRUE
            eos
              more := FALSE
  :

  PROC summarize([W]INT p)
    SEQ
      write(nl)
      write(nl)
      writeln("p =")
      writelong(p)
      write(nl)
      writeln("is a certified prime.")
      write(nl)
      write(nl)
  :

  PROC open(CHAN OF FILE inp, out)
    SEQ
      out!char;reallength
      out!int;12
      out!char;fraclength
      out!int;4
      out!char;intlength
      out!int;3
  :

  PROC close(VAL INT time, CHAN OF FILE inp, out)
    -- 1 second = 15625 units of 64 ms
    VAL units IS REAL64 ROUND time:
    VAL Tp IS units/15625.0(REAL64):
    VAL eps IS 0.00005(REAL64):
    SEQ
      out!char;nl
      out!str;18::"  p      N      Tp (s)"
      out!char;nl
      out!char;nl
```

```
      out!int;p
      out!str;2::"  "
      out!int;N
      out!str;1::" "
      out!real;Tp + eps
      out!char;nl
      out!char;nl
      out!str;10::"type eof: "
      inp?CASE eof
      out!eof
  :

  [W]INT x:
  INT t1, t2, seed0, bits:
  TIMER clock:
  CHAN OF LONG d:
  SEQ
     open(inp0, out0)
     select(seed0, bits)
     initialize(seed0)
     IF
       testing1
         longtest()
       TRUE
         SKIP
     IF
       testing2
         overflowtest()
       TRUE
         SKIP
     clock?t1
     PAR
       master(bits, x, d, a.out, b.inp, c.out)
       collector(a.inp, d)
     clock?t2
     summarize(x)
     close(t2 MINUS t1, inp0, out0)
  :

PROC ELEMENT(VAL INT i,
  CHAN OF LONG a.inp, a.out, c.inp, c.out, b.inp, b.out)

  PROC certifier(VAL INT id, CHAN OF LONG inp, out)

    PROC broadcast([W]INT n, BOOL more, VAL INT id, CHAN OF LONG inp, out)
      inp?CASE
        global;n
          SEQ
            IF
              id > 1
                out!global;n
              TRUE
                SKIP
            more := TRUE
        eos
          SEQ
            out!eos
            more := FALSE
    :

    PROC ShiftIn([W]INT q, BOOL more, VAL INT id, CHAN OF LONG inp, out)
      INT i:
      SEQ
        i := 1
```

```
      more := TRUE
      WHILE more AND (i < id)
        inp?CASE
          data;q
            SEQ
              out!data;q
              i := i+1
          complete
            SEQ
              out!complete
              more := FALSE
      IF
        more
          inp?CASE
            data;q
              SKIP
            complete
              SEQ
                out!complete
                more := FALSE
        TRUE
          SKIP
:

PROC ShiftOut(BOOL pass, more, VAL INT id, CHAN OF LONG inp, out)
  BOOL b:
  INT i:
  SEQ
    out!result;pass
    i := id+1
    more := TRUE
    WHILE more AND (i <= p)
      SEQ
        inp?CASE
          result;b
            SEQ
              out!result;b
              i := i+1
          complete
            SEQ
              out!complete
              more := FALSE
:

[W]INT n, q, n1, e, x:
BOOL pass, continue, more:
SEQ
  initialize()
  broadcast(n, continue, id, inp, out)
  WHILE continue
    SEQ
      n1 := n
      decrement(n1)
      more := TRUE
      WHILE more
        SEQ
          ShiftIn(q, more, id, inp, out)
          IF
            more
              SEQ
                x := two
                e := n1
                divide(e, q)
                modpower(x, e, n)
```

```
                        pass := NOT equal(x, one)
                        ShiftOut(pass, more, id, inp, out)
                  TRUE
                      SKIP
            broadcast(n, continue, id, inp, out)
  :

PROC generator(VAL INT id, CHAN OF LONG inp1, out1, inp2, out2)

  PROC witness([W]INT x, p, BOOL sure)
    [W]INT e, m, p1, r, y:
    SEQ
      m := one
      y := x
      e := p
      decrement(e)
      p1 := e
      sure := FALSE
      WHILE (NOT sure) AND greater(e, zero)
        IF
          odd(e)
            SEQ
              multiply(m, y)
              modulo(m, p)
              decrement(e)
          TRUE
            SEQ
              r := y
              square(y)
              modulo(y, p)
              halve(e)
              IF
                equal(y, one)
                  sure := less(one, r) AND less(r, p1)
                TRUE
                  SKIP
      sure := sure OR (NOT equal(m, one))
  :

  BOOL FUNCTION SmallFactor(VAL [W]INT a)
    INT i, j, len, p, carry, dummy:
    BOOL factor:
    VALOF
      SEQ
        j := 1
        factor := FALSE
        len := length(a)-1
        WHILE (NOT factor) AND (j < numprimes)
          SEQ
            p := primes[j]
            carry := 0
            i := len
            WHILE i >= 0
              SEQ
                dummy, carry := LONGDIV(carry, a[i], p)
                i := i-1
            factor := (carry = 0)
            j := j+1
      RESULT factor
  :

  PROC searcher(VAL INT id, CHAN OF LONG inp1, out1, out2)

    PROC ExtendProduct(VAL INT bits, [W]INT n, [numprimes]BOOL factor)
```

```
          [W]INT x:
          INT i:
          SEQ
            SEQ i = 0 FOR numprimes
              factor[i] := FALSE
            WHILE size(n) < bits
              SEQ
                randomdigit(i)
                i := i REM numprimes
                product(x, n, primes[i])
                IF
                  size(x) <= bits
                    SEQ
                      n := x
                      factor[i] := TRUE
                  TRUE
                    SKIP
:

PROC FindPrime(INT bits, [W]INT x, CHAN OF LONG inp1, out1, out2)
  [W]INT n, n1, y:
  [numprimes]BOOL factor:
  BOOL more, found:
  SEQ
    more := TRUE
    WHILE more
      SEQ
        out1!check
        inp1?CASE
          complete
            more := FALSE
          continue
            SEQ
              n := x
              ExtendProduct(bits, n, factor)
              increment(n)
              IF
                NOT SmallFactor(n)
                  SEQ
                    n1 := n
                    decrement(n1)
                    y := two
                    modpower(y, n1, n)
                    found := equal(y, one)
                TRUE
                  found := FALSE
              IF
                found
                  SEQ
                    out2!candidate;n;factor
                    out1!complete
                    more := FALSE
                TRUE
                  SKIP
    out2!complete
:

[W]INT x:
INT bits:
BOOL more:
SEQ
  more := TRUE
  WHILE more
    inp1?CASE
```

```
        start;bits;x
          FindPrime(bits, x, inp1, out1, out2)
        eos
          SEQ
            out2!eos
            more := FALSE
:

PROC reader(VAL INT id, CHAN OF LONG inp1, out1, inp2, out2)

  PROC monitor(VAL INT id, CHAN OF LONG inp1, out1, inp2, out2)
    BOOL more:
    SEQ
      more := TRUE
      WHILE more
        ALT
          inp1?CASE complete
            SEQ
              IF
                id < p
                  out1!complete
                TRUE
                  SKIP
              inp2?CASE
                check
                  out2!complete
                complete
                  SKIP
              more := FALSE
          inp2?CASE
            check
              out2!continue
            complete
              SEQ
                IF
                  id < p
                    out1!complete
                  TRUE
                    SKIP
                inp1?CASE complete
                more := FALSE
  :

  [W]INT x:
  INT bits:
  BOOL more:
  SEQ
    more := TRUE
    WHILE more
      inp1?CASE
        start;bits;x
          SEQ
            IF
              id < p
                out1!start;bits;x
              TRUE
                SKIP
            out2!start;bits;x
            monitor(id, inp1, out1, inp2, out2)
        eos
          SEQ
            IF
              id < p
                out1!eos
```

```
              TRUE
                SKIP
            out2!eos
            more := FALSE
:

PROC writer(VAL INT id, CHAN OF LONG inp1, inp2, out2)

  PROC PassResult(VAL INT terminate, CHAN OF LONG inp1, inp2, out2,
    BOOL done)
    [W]INT n:
    [numprimes]BOOL factor:
    INT count:
    BOOL passing:
    SEQ
      done := FALSE
      passing := TRUE
      count := 0
      WHILE (NOT done) AND (count < terminate)
        ALT
          inp1?CASE
            candidate;n;factor
              IF
                passing
                  SEQ
                    out2!candidate;n;factor
                    passing := FALSE
                TRUE
                  SKIP
            complete
              SEQ
                count := count+1
                IF
                  count = terminate
                    out2!complete
                  TRUE
                    SKIP
            eos
              done := TRUE
          inp2?CASE
            candidate;n;factor
              IF
                passing
                  SEQ
                    out2!candidate;n;factor
                    passing := FALSE
                TRUE
                  SKIP
            complete
              SEQ
                count := count+1
                IF
                  count = terminate
                    out2!complete
                  TRUE
                    SKIP
  :

  INT terminate:
  BOOL done:
  SEQ
    IF
      id < p
        terminate := 2
```

```
          TRUE
            terminate := 1
        done := FALSE
        WHILE NOT done
          PassResult(terminate, inp1, inp2, out2, done)
    :

    CHAN OF LONG a, b, c:

    SEQ
      initialize(((id*id)*id)
      PAR
        reader(id, inp1, out1, a, b)
        searcher(id, b, a, c)
        writer(id, c, inp2, out2)
  :

  PRI PAR
    certifier(i, a.inp, a.out)
    generator(i, c.inp, c.out, b.inp, b.out)
:

PROC FILES(CHAN OF ANY inp1, out1, CHAN OF FILE inp2, out2)
  #USE "/iolib/io.lib"
  IO(out1, inp1, out2, inp2)
:

CHAN OF ANY anyinp, anyout:
CHAN OF FILE fileinp, fileout:
[p + 1]CHAN OF LONG a, b, c:
PLACED PAR
  PROCESSOR 0 T8
    PLACE fileout AT inp0:
    PLACE fileinp AT out0:
    PLACE anyinp AT inp2:
    PLACE anyout AT out2:
    FILES(anyinp, anyout, fileout, fileinp)
  PROCESSOR 1 T8
    PLACE fileinp AT inp0:
    PLACE fileout AT out0:
    PLACE a[0] AT inp1:
    PLACE a[p] AT out2:
    PLACE b[0] AT inp3:
    PLACE c[0] AT out3:
    USER(fileinp, fileout, a[0], a[p], b[0], c[0])
  PLACED PAR i = 1 FOR p
    PROCESSOR i + 1 T8
      PLACE a[i] AT inp0:
      PLACE a[i-1] AT out1:
      PLACE c[i-1] AT inp2:
      PLACE b[i-1] AT out2:
      PLACE b[i] AT inp3:
      PLACE c[i] AT out3:
      ELEMENT(i, a[i], a[i-1], c[i-1], c[i], b[i], b[i-1])
```

Bibliography

[Adleman 87]   L. Adleman and M. Huang, "Recognizing primes in random polynomial time," *Proceedings of the 19th ACM STOC*, 462-469 (1987).

[Agnew 88]   G. Agnew, R. Mullin, and S. Vanstone, "Fast exponentiation in $GF(2^n)$," *Advances in Cryptology—EUROCRYPT '88*, Springer-Verlag Lecture Notes in Computer Science, 245-250 (1988).

[Ali 91]   A. Ali and C. Hartmann, "SIMPLE: A new approach to combinational circuit testing," *Proceedings of the Second Annual Symposium on Communications, Signal Processing, Expert Systems and ASIC VLSI Design* (1991).

[Bach 84]   E. Bach, *Analytic Methods in the Analysis and Design of Number-Theoretic Algorithms*, MIT Press, (1984).

[Bach 88]   E. Bach, "How to generate factored random numbers," *SIAM Journal on Computing*, **17**, 179-193 (1988).

[Barrett 86]   P. Barrett, "Implementing the Rivest Shamir and Adleman public key encryption algorithm on a standard digital signal processor," *Advances in Cryptology—CRYPTO '86*, Springer-Verlag Lecture Notes in Computer Science, 302-323 (1986).

[Beauchemin 86]   P. Beauchemin et al, "Two observations on probabilistic primality testing," *Advances in Cryptology—CRYPTO '86*, Springer-Verlag Lecture Notes in Computer Science, 443-450 (1986).

[Beth 86]   T. Beth, B. Cook, and D. Gollmann, "Architectures for exponentiation in $GF(2^n)$," *Advances in Cryptology—CRYPTO '86*, Springer-Verlag Lecture Notes in Computer Science, 302-310 (1986).

[Beth 91]   T. Beth, M. Frisch, and G. Simmons (Eds.), *Public-Key Cryptography: State of the Art and Future Directions*, E.I.S.S. Workshop Final Report, Springer-Verlag Lecture Notes in Computer Science (1991).

[Blakley 79]   B. Blakley and G. Blakley, "Security of number theoretic pub-
                       lic-key cryptosystems against random attack," *Cryptologia*; Part
                       I: **2**:4, 305-321 (1978); Part II: **3**:1, 29-42 (1979); Part III: **3**:2,
                       105-118 (1979).

[Bos 89]   J. Bos and M. Coster, "Addition chain heuristics," *Advances in
                       Cryptology—CRYPTO '89*, Springer-Verlag Lecture Notes in
                       Computer Science, 400-407 (1989).

[Brassard 88]   G. Brassard, *Modern Cryptology: A Tutorial*, Springer-Verlag
                       Lecture Notes in Computer Science (1988).

[Bressoud 89]   D. Bressoud, *Factorization and Primality Testing*, Springer-
                       Verlag (1989).

[Brickell 89]   E. Brickell, "A survey of hardware implementations of RSA,"
                       Advances in Cryptology—CRYPTO '89, Springer-Verlag
                       Lecture Notes in Computer Science, 368-370 (1989).

[Brinch Hansen 87]   P. Brinch Hansen, "Joyce—A programming language for dis-
                       tributed systems," *Software—Practice and Experience*, **17**, 29-
                       50 (1987).

[Brinch Hansen 89]   P. Brinch Hansen, "The Joyce language report," *Software—
                       Practice and Experience*, **19**, 553-578 (1989).

[Brinch Hansen 90a]   P. Brinch Hansen, "Householder reduction of linear equa-
                       tions," School of Computer and Information Science, Syracuse
                       University (1990). Revised version in *ACM Computing Sur-
                       veys*, **24**, 185-194 (1992).

[Brinch Hansen 90b]   P. Brinch Hansen, "The all-pairs pipeline," School of Com-
                       puter and Information Science, Syracuse University (1990).

[Brinch Hansen 90c]   P. Brinch Hansen, "Balancing a pipeline by folding," School of
                       Computer and Information Science, Syracuse University
                       (1990).

[Brinch Hansen 91a]   P. Brinch Hansen, "The n-body pipeline," School of Computer
                       and Information Science, Syracuse University (1991).

[Brinch Hansen 91b]   P. Brinch Hansen, "A generic multiplication pipeline," School
                       of Computer and Information Science, Syracuse University
                       (1991).

[Brinch Hansen 91c]   P. Brinch Hansen, "The fast Fourier transform," School of
                       Computer and Information Science, Syracuse University
                       (1991).

[Brinch Hansen 91d] P. Brinch Hansen, "Parallel divide and conquer," School of Computer and Information Science, Syracuse University (1991).

[Brinch Hansen 91e] P. Brinch Hansen, "Do hypercubes sort faster than tree machines?," School of Computer and Information Science, Syracuse University (1991). Also in *Concurrency—Practice and Experience*, **6**, 143-151 (1994).

[Brinch Hansen 92a] P. Brinch Hansen, "Simulated annealing," School of Computer and Information Science, Syracuse University (1992).

[Brinch Hansen 92b] P. Brinch Hansen, "Primality testing," School of Computer and Information Science, Syracuse University (1992).

[Brinch Hansen 92c] P. Brinch Hansen, "Parallel Monte Carlo trials," School of Computer and Information Science, Syracuse University (1992).

[Brinch Hansen 92d] P. Brinch Hansen, "Multiple-length division revisited: A tour of the minefield," School of Computer and Information Science, Syracuse University (1992). Revised version in *Software—Practice and Experience*, **24**, 579-601 (1994).

[Brinch Hansen 92e] P. Brinch Hansen, "Numerical solution of Laplace's equation," School of Computer and Information Science, Syracuse University (1992).

[Brinch Hansen 92f] P. Brinch Hansen, "Parallel cellular automata: A model program for computational science," School of Computer and Information Science, Syracuse University (1992). Invited paper for *Concurrency—Practice and Experience*, **5**, 425-448 (1993).

[Brinch Hansen 93] P. Brinch Hansen, "Model programs for computational science: A programming methodology for multicomputers," School of Computer and Information Science, Syracuse University (1993). Invited paper for *Concurrency—Practice and Experience*, **5**, 407-423 (1993).

[Cok 91] R. Cok, *Parallel Programs for the Transputer*, Prentice Hall (1991).

[Cole 89] M. Cole, *Algorithmic Skeletons: Structured Management of Parallel Computation*, MIT Press (1989).

[Cooper 90] R. Cooper and W. Patterson, "RSA as a benchmark for multiprocessor machines," *AUSCRYPT '90*, Springer-Verlag Lecture Notes in Computer Science, 356-359 (1990).

[Denning 82] D. Denning, *Cryptography and Data Security*, Addison-Wesley (1982).

[Diffie 76]   W. Diffie and M. Hellman, "New directions in cryptography,"
                    IEEE Transactions on Information Theory, **IT-22**, 644-654
                    (1976).

[Er 91]   M. Err et al, "Design and implementation of an RSA cryptosys-
                    tem using multiple DSP chips," *Microprocessors and Microsys-
                    tems*, **15**:7, 369-378 (1991).

[Findlay 89]   P. Findlay and B. Johnson, "Modular exponentiation using re-
                    cursive sums of residues," *Advances in Cryptology—CRYPTO
                    '89*, Springer-Verlag Lecture Notes in Computer Science, 371-
                    386 (1989).

[Fox 88]   G. Fox et al, *Solving Problems on Concurrent Processors*,
                    volume 1, Prentice Hall (1988).

[Geiselmann 90]   W. Geiselmann and D. Gollmann, "VLSI design for exponen-
                    tiation in $GF(2^n)$," *Advances in Cryptology—AUSCRYPT
                    '90*, Springer-Verlag Lecture Notes in Computer Science, 398-
                    405 (1990).

[Ghafoor 89]   A. Ghafoor and A. Singh, "Systolic architecture for finite field
                    exponentiation," *IEE Proceedings*, **136**, Part E, 465-470
                    (1989).

[Goldwasser 86]   S. Goldwasser and J. Kilian, "Almost all primes can be quickly
                    certified," *Proceedings of the 18th ACM STOC*, 316-329
                    (1986).

[Gordon 84]   J. Gordon, "Strong primes are easy to find," *Advances in
                    Cryptology—EUROCRYPT '84*, Springer-Verlag Lecture
                    Notes in Computer Science, 216-223 (1984).

[Hartmann 91]   C. Hartmann and D. Shiau, "Digital test generation using mul-
                    tiprocessing," School of Computer and Information Science,
                    Syracuse University (1991).

[Hastad 85]   J. Hastad, "On using RSA with low exponent in a public key
                    network," *Advances in Cryptology—CRYPTO '85*, Springer-
                    Verlag Lecture Notes in Computer Science, 403-408 (1985).

[Hoornaert 88]   F. Hoornaert et al, "Fast RSA-hardware: Dream or reality?,"
                    Advances in Cryptology—EUROCRYPT '88, Springer-Verlag
                    Lecture Notes in Computer Science, 257-264 (1988).

[INMOS 88a]   INMOS, Ltd., IMS T800 transputer (1988).

[INMOS 88b]   INMOS, Ltd., *The occam 2 Reference Manual*, Prentice-Hall
                    International (1988).

[Jung 87]   A. Jung, "Implementing the RSA cryptosystem," *Computers and Security*, **6**, 342-350 (1987).

[Kawamura 88]   S. Kawamura and K. Hirano, "A fast modular arithmetic algorithm using a residue table," *Advances in Cryptology—EUROCRYPT '88*, Springer-Verlag Lecture Notes in Computer Science, 245-250 (1988).

[Knuth 82]   D. Knuth, *The Art of Computer Programming, Volume 2: Seminumerical Algorithms*, 2nd Edition, Addison-Wesley (1982).

[Lenstra 87]   A. Lenstra and H. Lenstra, Jr., "Algorithms in number theory," University of Chicago, Department of Computer Science, Technical Report #87-008 (1987).

[Maurer 89]   U. Maurer, "Fast generation of secure RSA-moduli with almost maximal diversity," *Advances in Cryptology—EUROCRYPT '89*, Springer-Verlag Lecture Notes in Computer Science, 636-647 (1989).

[Meiko 88]   Meiko, Ltd., *Occam Programming System User Manual* (1988).

[Miller 76]   G. Miller, "Riemann's hypothesis and tests for primality," *Journal of Computer and System Sciences*, **13**, 300-317 (1976).

[Morita 89]   H. Morita, "A fast modular multiplication algorithm based on a higher radix," *Advances in Cryptology—CRYPTO '89*, Springer-Verlag Lecture Notes in Computer Science, 387-399 (1989).

[Natarajan 89]   K. Natarajan, "Expected performance of parallel search," *1989 International Conference on Parallel Processing, vol. III*, 121-125 (1989).

[Nelson 87]   P. Nelson, *Parallel Programming Paradigms*, Ph.D. Dissertation, University of Washington, Seattle, Washington (1987).

[Niven 80]   I. Niven and H. Zuckerman, *An Introduction to The Theory of Numbers*, fourth edition, John Wiley & Sons (1980).

[Orton 86]   G. Orton et al, "VLSI implementation of public-key encryption algorithms," *Advances in Cryptology—CRYPTO '86*, Springer-Verlag Lecture Notes in Computer Science, 277-301 (1986).

[Orup 90]   H. Orup, E. Svendsen, and E. Adreasen, "VICTOR—an effi-
                    cient RSA hardware implementation," *Advances in Cryptol-
                    ogy—EUROCRYPT '90*, Springer-Verlag Lecture Notes in
                    Computer Science, 245-252 (1990).

[Pollard 74]   J. Pollard, "Theorems on factorization and primality testing,"
                    Proc. Cambr. Philos. Society, **76**, 521-528 (1974).

[Pollard 75]   J. Pollard, "A Monte Carlo method for factorization," *BIT*, **15**,
                    331-334 (1975).

[Pomerance 87]   C. Pomerance, "Very short primality proofs," *Mathematics of
                    Computation*, **48**, 315-322 (1987).

[Pratt 75]   V. Pratt, "Every prime has a succint certificate," *SIAM Journal
                    on Computing*, **4**, 214-220 (1975).

[Rabin 80]   M. Rabin, "Probabilistic Algorithm for Testing Primality,"
                    Journal of Number Theory, **12**, 128-138 (1980).

[Rao 91]   M. Rao, *Performance Efficient Parallel Programming*, Ph.D.
                    Dissertation, Carnegie-Mellon University, Pittsburgh, Pennsyl-
                    vania (1991).

[Rankine 86]   G. Rankine, "A complete single chip RSA device," *Advances
                    in Cryptology—CRYPTO '86*, Springer-Verlag Lecture Notes
                    in Computer Science, 480-487 (1986).

[Ribenboim 91]   P. Ribenboim, *The Little Book of Big Primes*, Springer-Verlag
                    (1991).

[Riesel 85]   H. Riesel, *Prime Numbers and Computer Methods for Factor-
                    ization*, Birkhauser (1985).

[Rivest 78a]   R. Rivest, "Remarks on a proposed cryptanalytic attack on the
                    M.I.T. public-key cryptosystem," *Cryptologia*, 2:1, 62-65
                    (1978).

[Rivest 78b]   R. Rivest, A. Shamir and L. Adleman, "A method for obtaining
                    digital signatures and public-key cryptosystems," *Communica-
                    tions of the ACM*, **21**:2, 120-126 (1978).

[Rivest 80]   R. Rivest, "A description of a single-chip implementation of the
                    RSA cipher," *LAMBDA Magazine*, **1**:3, 14-18 (1980).

[Rivest 90]   R. Rivest, "Finding four million large random primes," *Ad-
                    vances in Cryptology—CRYPTO '90*, Springer-Verlag Lecture
                    Notes in Computer Science, 625-626 (1990).

[Shawe-Taylor 86]   J. Shawe-Taylor, "Generating strong primes," *Electronics Let-
                    ters*, **22**, 875-877 (1986).

[Silverman 91]   R. Silverman, "Massively distributed computing and factoring large integers," *Communications of the ACM*, **34**:11, 95-103 (1991).

[Stanford-Clark 91]   A. Stanford-Clark, *Parallel Paradigms and Their Implementation*, Ph.D. Dissertation, University of East Anglia, United Kindom (1991).

[Welsh 88]   D. Welsh, *Codes and Cryptography*, Oxford University Press (1988).

[Williams 82]   H. Williams, "A p+1 method of factoring," *Mathematics of Computation*, **39**, 225-234 (1982).

[Wirth 71]   N. Wirth, "The programming language Pascal," *Acta Informatica*, **1**, 35-63 (1971).

[Yacobi 90]   Y. Yacobi, "Exponentiating faster with addition chains" *Advances in Cryptology—EUROCRYPT '90*, Springer-Verlag Lecture Notes in Computer Science, 222-229 (1990).

[Yelick 93]   K. Yelick, "Programming models for irregular applications," *SIGPLAN Notices*, **28**:1, 28-31 (1993).

[Yun 90]   D. Yun and C. Zhang, "Parallel algorithms and designs for fundamental operations in cryptography," in G. Zobrist (Ed.), *Progress in Computer-Aided VLSI Design*, volume 3, 1-37 (1990).

Index

List of Figures

List of Tables

List of Algorithms

Lecture Notes in Computer Science

For information about Vols. 1–792
please contact your bookseller or Springer-Verlag

Vol. 830: C. Castelfranchi, E. Werner (Eds.), Artificial Social Systems. Proceedings, 1992. XVIII, 337 pages. 1994. (Subseries LNAI).

Vol. 831: V. Bouchitté, M. Morvan (Eds.), Orders, Algorithms, and Applications. Proceedings, 1994. IX, 204 pages. 1994.

Vol. 832: E. Börger, Y. Gurevich, K. Meinke (Eds.), Computer Science Logic. Proceedings, 1993. VIII, 336 pages. 1994.

Vol. 833: D. Driankov, P. W. Eklund, A. Ralescu (Eds.), Fuzzy Logic and Fuzzy Control. Proceedings, 1991. XII, 157 pages. 1994. (Subseries LNAI).

Vol. 834: D.-Z. Du, X.-S. Zhang (Eds.), Algorithms and Computation. Proceedings, 1994. XIII, 687 pages. 1994.

Vol. 835: W. M. Tepfenhart, J. P. Dick, J. F. Sowa (Eds.), Conceptual Structures: Current Practices. Proceedings, 1994. VIII, 331 pages. 1994. (Subseries LNAI).

Vol. 836: B. Jonsson, J. Parrow (Eds.), CONCUR '94: Concurrency Theory. Proceedings, 1994. IX, 529 pages. 1994.

Vol. 837: S. Wess, K.-D. Althoff, M. M. Richter (Eds.), Topics in Case-Based Reasoning. Proceedings, 1993. IX, 471 pages. 1994. (Subseries LNAI).

Vol. 838: C. MacNish, D. Pearce, L. Moniz Pereira (Eds.), Logics in Artificial Intelligence. Proceedings, 1994. IX, 413 pages. 1994. (Subseries LNAI).

Vol. 839: Y. G. Desmedt (Ed.), Advances in Cryptology - CRYPTO '94. Proceedings, 1994. XII, 439 pages. 1994.

Vol. 840: G. Reinelt, The Traveling Salesman. VIII, 223 pages. 1994.

Vol. 841: I. Prívara, B. Rovan, P. Ružička (Eds.), Mathematical Foundations of Computer Science 1994. Proceedings, 1994. X, 628 pages. 1994.

Vol. 842: T. Kloks, Treewidth. IX, 209 pages. 1994.

Vol. 843: A. Szepietowski, Turing Machines with Sublogarithmic Space. VIII, 115 pages. 1994.

Vol. 844: M. Hermenegildo, J. Penjam (Eds.), Programming Language Implementation and Logic Programming. Proceedings, 1994. XII, 469 pages. 1994.

Vol. 845: J.-P. Jouannaud (Ed.), Constraints in Computational Logics. Proceedings, 1994. VIII, 367 pages. 1994.

Vol. 846: D. Shepherd, G. Blair, G. Coulson, N. Davies, F. Garcia (Eds.), Network and Operating System Support for Digital Audio and Video. Proceedings, 1993. VIII, 269 pages. 1994.

Vol. 847: A. L. Ralescu (Ed.) Fuzzy Logic in Artificial Intelligence. Proceedings, 1993. VII, 128 pages. 1994. (Subseries LNAI).

Vol. 848: A. R. Krommer, C. W. Ueberhuber, Numerical Integration on Advanced Computer Systems. XIII, 341 pages. 1994.

Vol. 849: R. W. Hartenstein, M. Z. Servít (Eds.), Field-Programmable Logic. Proceedings, 1994. XI, 434 pages. 1994.

Vol. 850: G. Levi, M. Rodríguez-Artalejo (Eds.), Algebraic and Logic Programming. Proceedings, 1994. VIII, 304 pages. 1994.

Vol. 851: H.-J. Kugler, A. Mullery, N. Niebert (Eds.), Towards a Pan-European Telecommunication Service Infrastructure. Proceedings, 1994. XIII, 582 pages. 1994.

Vol. 852: K. Echtle, D. Hammer, D. Powell (Eds.), Dependable Computing – EDCC-1. Proceedings, 1994. XVII, 618 pages. 1994.

Vol. 853: K. Bolding, L. Snyder (Eds.), Parallel Computer Routing and Communication. Proceedings, 1994. IX, 317 pages. 1994.

Vol. 854: B. Buchberger, J. Volkert (Eds.), Parallel Processing: CONPAR 94 – VAPP VI. Proceedings, 1994. XVI, 893 pages. 1994.

Vol. 855: J. van Leeuwen (Ed.), Algorithms – ESA '94. Proceedings, 1994. X, 510 pages.1994.

Vol. 856: D. Karagiannis (Ed.), Database and Expert Systems Applications. Proceedings, 1994. XVII, 807 pages. 1994.

Vol. 857: G. Tel, P. Vitányi (Eds.), Distributed Algorithms. Proceedings, 1994. X, 370 pages. 1994.

Vol. 858: E. Bertino, S. Urban (Eds.), Object-Oriented Methodologies and Systems. Proceedings, 1994. X, 386 pages. 1994.

Vol. 859: T. F. Melham, J. Camilleri (Eds.), Higher Order Logic Theorem Proving and Its Applications. Proceedings, 1994. IX, 470 pages. 1994.

Vol. 860: W. L. Zagler, G. Busby, R. R. Wagner (Eds.), Computers for Handicapped Persons. Proceedings, 1994. XX, 625 pages. 1994.

Vol: 861: B. Nebel, L. Dreschler-Fischer (Eds.), KI-94: Advances in Artificial Intelligence. Proceedings, 1994. IX, 401 pages. 1994. (Subseries LNAI).

Vol. 862: R. C. Carrasco, J. Oncina (Eds.), Grammatical Inference and Applications. Proceedings, 1994. VIII, 290 pages. 1994. (Subseries LNAI).

Vol. 863: H. Langmaack, W.-P. de Roever, J. Vytopil (Eds.), Formal Techniques in Real-Time and Fault-Tolerant Systems. Proceedings, 1994. XIV, 787 pages. 1994.

Vol. 864: B. Le Charlier (Ed.), Static Analysis. Proceedings, 1994. XII, 465 pages. 1994.

Vol. 865: T. C. Fogarty (Ed.), Evolutionary Computing. Proceedings, 1994. XII, 332 pages. 1994.

Vol. 866: Y. Davidor, H.-P. Schwefel, R. Männer (Eds.), Parallel Problem Solving from Nature-Evolutionary Computation. Proceedings, 1994. XV, 642 pages. 1994.

Vol 867: L. Steels, G. Schreiber, W. Van de Velde (Eds.), A Future for Knowledge Acquisition. Proceedings, 1994. XII, 414 pages. 1994. (Subseries LNAI).

Vol. 868: R. Steinmetz (Ed.), Multimedia: Advanced Teleservices and High-Speed Communication Architectures. Proceedings, 1994. IX, 451 pages. 1994.

Vol. 869: Z. W. Raś, Zemankova (Eds.), Methodologies for Intelligent Systems. Proceedings, 1994. X, 613 pages. 1994. (Subseries LNAI).

Vol. 870: J. S. Greenfield, Distributed Programming Paradigms with Cryptography Applications. XI, 182 pages. 1994.

Vol. 871: J. P. Lee, G. G. Grinstein (Eds.), Database Issues for Data Visualization. Proceedings, 1993. XIV, 229 pages. 1994.